THE OSCILLATING BRAIN

HOW OUR BRAIN WORKS

TIMOTHY D. SHEEHAN, M.D.

LifeRich PUBLISHING

LifeRich Publishing is a registered trademark of The Reader's Digest Association, Inc.

LifeRich Publishing books may be ordered through booksellers or by contacting:

LifeRich Publishing
1663 Liberty Drive
Bloomington, IN 47403
www.liferichpublishing.com
1 (888) 238-8637

ISBN: 978-1-4897-0580-8 (sc)
ISBN: 978-1-4897-0581-5 (hc)
ISBN: 978-1-4897-0582-2 (e)

Library of Congress Control Number: 2015920839

Print information available on the last page.

LifeRich Publishing rev. date: 1/12/2016

To my wife, Irma, and my children, Tim and Lisa

CONTENTS ————————————————

PREFACE ─────────────────────────────────

I'm a psychiatrist. Psychiatrists are physicians who specialize in the treatment of mental disorders. I entered psychiatry hoping to understand human behavior. I was looking for a field that encompassed both the biological and psychological aspects of human behavior. I believe that understanding human behavior requires understanding brain function. I've spent my professional lifetime trying to sort out how the brain works. This book presents an explanation of brain function that's consistent with my experience and the findings of the neuroscience community.

It wouldn't have crossed my mind to go into psychiatry during childhood and adolescence. My parents were devout Irish Catholics. After attending Catholic elementary and high schools, my only goal was to enter the Society of Jesus—a Catholic religious order more commonly known as the Jesuits.

I graduated from high school in 1962 and spent the next two years in the Jesuits. The initial two years of Jesuit training are referred to as the novitiate—a time of prayerful transformation into a soldier of Christ. Life in the novitiate was highly regulated and tightly focused on developing one's spiritual life. In some ways, it was a bit like army basic training, which also strives to transform civilians into soldiers.

To my surprise, I began experiencing religious doubts during the second year of my novitiate training. As you can readily imagine, religious doubts are not compatible with life in the Jesuits. I left the Jesuits in 1964, shortly before completing the novitiate. I had no idea what I would do next. I'd never had a plan B.

I began a slow process of reexamining everything I'd ever taken for granted. It's been an extremely personal process. Sharing my existential angst seemed pointless. I found myself in a world without the certainties of the religious faith passed down by my parents, and I didn't trust anyone to sort it out for me. I felt a need to find the answers myself.

College was an obvious choice. In college, I was surrounded by people with no idea of where they were going. I gravitated toward an interest in human behavior and majored in psychology. However, psychology didn't appear to provide the type of answers I was seeking. I came away with a feeling that psychology describes, without actually explaining, human behavior. I had no interest in pursuing postgraduate training in psychology.

I graduated from college in 1968, still without a plan B. However, I didn't have to wonder what would happen next. The Vietnam War was in full force, and I expected to be drafted shortly after dropping my school deferment.

Rather than wait for a draft notice, I entered army active duty in December 1968 under the Officer Candidate School (OCS) option. At that time, the army OCS option required nearly a year of training before being commissioned a second lieutenant: approximately two months of basic combat training; two months of Advanced Individual Training (AIT), where I learned entry-level skills of an enlisted combat engineer; and six months of Infantry OCS at Fort Benning, Georgia, where I learned the skills required of an infantry platoon leader. If I failed to complete the OCS program, AIT training enabled me to go directly to an enlisted combat engineer assignment. Enlisted combat engineers are always needed in combat zones.

Before completing Infantry OCS in November 1969, I'd finally decided on a plan B. Apparently, fresh air and exercise are conducive to making life decisions. I'd go to medical school and become a psychiatrist. True, I knew next to nothing about psychiatry, although it seemed like a field that would lead to an understanding of human behavior in terms of brain function.

Of course, plan B would have to wait for me to complete my military obligation. In 1969, the army was selecting a few graduates from each Infantry OCS class to serve as medical service corps officers. Medical service corps officers work in medical settings providing medical administrative support. That seemed more in line with my newly formed goal. I applied and was accepted for a commission as a medical service corps officer. Later, I realized that although I'd been trained to serve as an infantry platoon leader, I had surprisingly little training for the duties of a medical service corps officer.

Initially, I was assigned to Fort Sam Houston in San Antonio, Texas, as the chief of Training Support Branch at the training center for combat medics. Training Support Branch maintained and supplied the equipment needed for combat medic training. The branch was responsible for the rifles, radios, projectors, training films, medical equipment (including sterilizers), and even the fifty state flags (displayed at graduation parades). The operation involved about fifty enlisted soldiers, including seven noncommissioned officers (sergeants). I was the only person in the branch whose job wasn't clearly defined. I came away with the impression that the army relied heavily upon a sink-or-swim approach for career development of young medical service corps officers.

Following my assignment to Fort Sam Houston, I deployed to Vietnam. Prior to deployment, I married my wife, Irma. In Vietnam, I was assigned to the 1/327th Infantry Battalion of the 101st Airborne Division as one of two officers in the battalion medical platoon. The other officer was a physician, the battalion surgeon. Although I was addressed as "Doc" and asked medical questions by infantry officers, I had less medical training than an enlisted combat medic.

In August 1971, I returned from Vietnam and left active duty to pursue plan B. I rejoined my wife in San Antonio and spent the following year completing undergraduate premed requirements. In the process, I learned that I was eligible for an active-duty military scholarship program that would cover my medical school expenses while paying my active-duty salary. As a young married fellow expecting to have children, I found that attractive. I even looked forward to the possibility of performing a medical job in the army that I'd been trained for.

I returned to active duty in August 1972 when I started medical school at Georgetown University Medical School. Georgetown has an excellent reputation and I felt fortunate to be accepted there. I also hoped to transition directly to the army psychiatry residency training program at Walter Reed Army Medical Center since both are located in the Washington, DC, area.

After completing my medical training at Georgetown University Medical School in May 1976, I spent the next four years completing psychiatry training at Walter Reed Army Medical Center. Irma and I had our first child, Tim, while I was in medical school. We had our second child, Lisa, during my residency. I completed my psychiatry residency in 1980.

Psychiatrists are exposed to a wide range of abnormal human behavior. Psychiatric training is hospital-based and their patients often have severe psychiatric disorders, such as schizophrenia, bipolar disorder, and major depressive disorder. Psychiatrists also assist other physicians in the treatment of patients suffering from behavioral and cognitive changes associated with medical disorders, such as cancer, endocrine disease, heart disease, and medication-related changes. Medical disorders frequently interfere with brain function.

Psychiatrists are also well positioned to observe normal human behavior. Our patients often behave quite normally. The challenge lies in recognizing abnormal behavior when it occurs.

Human behavior exists across a wide spectrum in which "normal" and "abnormal" aren't always easy to distinguish.

Direct exposure to the spectrum of human behavior forced me to discard every preconceived idea I'd ever had about brain function. Along with generations of psychiatrists, I've struggled to find an explanation for the behavior that psychiatrists encounter. Fortunately, the neuroscience community has made significant advances in understanding brain operation. Drawing upon their work and recent work on complex systems, I've been able to formulate the explanation of human behavior based on brain function that is presented in this book.

I retired from the army in late 2000 after serving on active duty for nearly thirty-one years. During those years, I served in a wide variety of clinical and administrative settings. Since retiring from active duty, I've continued to work as an army psychiatrist in a civilian capacity for over fourteen years.

Army psychiatrists see a wide range of military personnel (active-duty and retired) and their dependents. Their patient population isn't limited by insurance coverage issues or focused on patients whose psychiatric disorders are so severe that they're compelled by family members (or the court) to seek treatment. Some soldiers and dependents do experience severe psychiatric disorders, such as schizophrenia, bipolar disorder, and major depressive disorder and may require psychiatric hospitalization. However, most are experiencing situational difficulties that don't require hospitalization.

Soldiers are exposed to a variety of stressors, such as separation from loved ones, problems with supervisors, and injuries due to military training or combat. New recruits may simply be unable to tolerate military regimentation and have to be returned to civilian status. Post-traumatic stress disorder, of course, is a frequent finding among combat veterans. Army psychiatrists also encounter a wide range of psychiatric disorders among military dependents. It's difficult to maintain a marriage and family when your spouse is away, possibly in harm's way, for extended periods.

In addition to patient care, army psychiatrists are continually interacting with a wide variety of complex systems. The military itself is a complex bureaucratic system. Medical units, line units, and military families are all complex organizational systems. Army psychiatrists are often required to play a leadership role, such as a clinic chief, in the army's medical system. They also have to be aware of organizational factors that impact their patients. Treatment often involves intervention with the soldier's unit or awareness of family dynamics that impact the soldier.

Over the years, I've become increasingly interested in the features of complex systems. As I've learned more about complex systems, I've come to appreciate that the principles that underlie complex systems provide a window of insight into brain function. The brain, of course, is a complex system. The principles of complex systems underlie both normal and abnormal brain function. I'll refer often to the principles of complex systems throughout my discussion of brain function. In the book's final section, I'll discuss the features of complex systems in more detail.

My understanding of human behavior owes a great deal to my wife and children. As you'll see, I place special emphasis on the importance of relationships, which has evolved directly from my relationship with my wife and children. Through those relationships, the expression "Love thy neighbor as thyself" has taken on concrete meaning. My experience in the military has extended my appreciation of relationships beyond my immediate family. I find that I'm on the firmest footing in my work as a psychiatrist when I treat others as I would want a member of my own family treated.

I'd like to give a special thanks to my sister, Margie. Margie has served as a sounding board and informal editor throughout this project. My goal has been to write a book that Margie finds readable. To the extent that you're able to easily follow the book's conceptual development, Margie deserves much of the credit. Hopefully, you'll be patient with me and persevere through any sections of the book where she's been less successful. Even Margie has limitations.

INTRODUCTION ——————————————

We've traditionally viewed the human brain as a black box—a system that can be approached only in terms of input and output without actually understanding how it works. We've focused on the stimuli we're exposed to, our subjective reactions, and our observed responses without pretending to understand the internal workings of the brain's three-pound mass of motionless gray matter.

Without an understanding of the brain's actual function, behavioral health specialists have gone in a variety of directions. Psychiatrists, for example, have traditionally valued their subjective reactions in response to a patient. My subjective reactions provide an additional source of information regarding a patient's relational difficulties. My reactions help me to gain insight into why the patient is experiencing difficulty in relationships outside of my office. Psychologists, on the other hand, tend to emphasize the importance of objective information. Psychological testing enables them to assess different aspects of personality in a more objective fashion. Within both psychiatry and psychology, different schools of thought have evolved that emphasize one or another aspect of human experience.

I'm reminded of the story of blind men describing an elephant. Their descriptions vary depending upon the specific area of the elephant they're in contact with. Even a carefully compiled summary of their descriptions is more likely to result in confusion than an accurate understanding of the elephant.

To understand human behavior, we need to understand how

the brain works. After all, human behavior is a product of brain function. Reality (the world as we experience it) is constructed by the brain. Reality, as our brains construct it, is inherently oversimplified and heavily biased by our emotional reactions. A better understanding of brain function can help us to be a bit less dogmatic in our views (the word *humility* comes to mind) and more receptive to nuanced (diplomatic) solutions to human differences.

This book provides an explanation of brain function. It brings together observations of clinical psychiatrists, neuroscientists, and complex-systems theorists into an explanatory synthesis. I realize that the explanation of brain function presented in this book isn't definitive. I doubt that a definitive explanation of brain function is currently feasible. After all, the body of neuroscience information is expanding exponentially and we're only beginning to understand the principles that govern complex systems. Nonetheless, I believe that the general explanation of brain function found in this book will stand up well to future investigation.

The book is written for anyone interested in human behavior and brain function. I've drawn upon the findings of neuroscientists and complex-systems theorists along with my own experience in clinical psychiatry. However, I'm writing for readers without a preexisting background in these fields.

Dr. Dale Purves, in his book *Brains: How They Seem to Work*, indicates that the "conception of how brains work has not been substantiated despite an effort that now spans 50 years."[1] Dr. Purves is a distinguished neuroscientist. He suggests that part of the reason for delay is "the absence of some guiding principle or principles that would help to understand the neural underpinnings of perceptual, behavioral, and cognitive phenomenology in a more general way."[2]

I believe that the guiding principles needed to understand brain function are those that underlie complex systems. Complex systems are patterns of activity that involve multiple variables and repetitive patterns of interaction. In the case of the brain, "repetitive patterns of interaction" refers to neural network oscillation—a central theme of this book. Thanks to the meticulous work of neuroscientists, such as Dr. Purves, our understanding of brain function has advanced to the point that I can offer a general explanation of brain function that integrates the experience of clinical psychiatry with the growing body of neuroscience information and our evolving understanding of complex systems. Throughout the book, I'll be referring to the principles of complex systems as they relate to brain function. I'll devote the book's final section to a more detailed discussion of complex systems.

I'd like to start by giving you an overview of the explanation of brain function found in this book. If I use terms that you're unfamiliar with, don't worry. Everything mentioned in this introductory section will be discussed in more detail as the book progresses.

Nerve cells (neurons) are the workhorses of brain activity. When a nerve cell is stimulated, electrical activity is transmitted along the neuron's outer membrane and down its axon. The axon is a long extension of the nerve cell that extends to other areas of the brain or body.

Neural activity is transferred from one nerve cell to another through neurotransmitters. Neurotransmitters are relatively simple chemicals released by neurons at tips of their axons. Neurotransmitters bridge the narrow gap (referred to as a synapse) that separates the axon tip of one neuron from the dendrites (short extensions) of the next neuron. The neuron releasing neurotransmitters is referred to as presynaptic. Neurotransmitters cross the synaptic gap and come in contact with the outer surface membrane of the postsynaptic neuron (the receiving neuron on the other side of the synapse). Dendrites of the postsynaptic neuron have neuroreceptors (proteins) embedded in their outer surface membrane. Neurotransmitters released by the presynaptic neuron interact with neuroreceptors of the postsynaptic neuron, influencing the postsynaptic neuron's level of activation. The interaction of presynaptic neurotransmitters with postsynaptic neuroreceptors enables neural activation to be transmitted from one neuron to another.

While transmission within a neuron is electrical in nature, transmission between neurons is chemical. This is an important distinction. Brain function requires the speed of electrical transmission to react rapidly to stimulation on a moment-to-moment basis. The speed of electrical transmission permits separate areas of the brain to work together almost instantaneously. Chemical transmission between neurons, on the other hand, permits neural network transformation in response to experience, a process essential for forming memories and modifying behavior based upon past experience.

Neurologists record the brain's electrical activity using an electroencephalograph. An electroencephalograph records the brain's electrical activity through sensors placed on the scalp. The recording it produces is called an electroencephalogram (EEG). Neurologists use EEGs to monitor normal brain activity and to look for evidence of abnormality, such as seizure activity.

Continuous electrical activity is an essential feature of the living brain. Regardless of our state of awareness, an electroencephalogram will show the wave patterns of the brain's ongoing electrical activity. Many of us have had an electrocardiogram (ECG) at one time or another. An electrocardiogram records the electrical activity of the heart through sensors placed on the chest. In death, our heart is no longer functioning and the lines of the ECG are flat. In brain death, the lines of the electroencephalogram are flat.

Neural network oscillation is responsible for the brain's sustained electrical activity. Neural oscillation is a form of repetitive activity. As I'll repeat frequently, repetitive activity is characteristic of complex systems. An individual nerve cell (neuron) transmits information in only one direction. Within the brain, however, communication is routinely bidirectional. For a unidirectional neuron to function in a bidirectional fashion, neural activity is directed back to a previously activated neuron. Reactivation of a previously fired neuron initiates a repetitive cycle of firing within a neuron loop involving a relatively small number of neurons. Within a neuron loop, a stable pattern of repetitive firing can be sustained indefinitely. The pattern of repetitive firing within such a loop of neurons is referred to as neural oscillation.

Neural network oscillation is the product of reciprocal connections. Reciprocal connections are characteristic of brain architecture. The brain's extensive reciprocal connections permit a variety of patterns of neural oscillation. While some patterns of neural oscillation are localized (restricted to specific regions of the brain), others involve reciprocal connections between geographically separate areas of the brain permitting more widespread neural network interaction. Neural network oscillation transforms the brain's three-pound mass of gray matter into a dynamic medium.

Sensory information is initially processed in specific local areas of the brain. Vision, for example, has its own visual processing area while hearing and somatic (bodily) perception each have independent processing areas. These specialized processing areas respond locally to visual, auditory, or somatic (bodily) sensations.

Conscious experience, on the other hand, involves the integrated interaction of the localized sensory processing areas of the brain and separately located areas of the brain responsible for directing motor activity. I use the word "global" to refer to this type of widespread brain activity. A globalized pattern of neural network oscillation (involving both the sensory and motor areas of the cerebral cortex) enables the brain to integrate sensory information into our experience of reality almost instantaneously allowing us to respond rapidly to events in a fashion that promotes survival.

Neurotransmitters permit the brain to maintain a baseline

frequency of cortical oscillation. Rapid-acting neurotransmitters permit sustained oscillatory interaction within a loop of repetitively firing neurons. Slower-acting neurotransmitters are responsible for modifying the oscillatory rate—locally (in response to sensory stimuli) and globally (as required for conscious perception). Both rapid-acting and slower-acting neurotransmitters are relatively simple molecules that nerve cells can produce in quantity and store in the vacuoles (spherical packets) in their axon tips.

By releasing neurotransmitter molecules into the synaptic gap, neurons are able to stimulate or inhibit the firing of other neurons. Rapid-acting neurotransmitters can be either activating or inhibiting. By varying the pattern of stimulation versus inhibition, the frequency of neural network oscillation is modified. Slower-acting neurotransmitters permit the oscillatory patterns of past neural network activity to influence current and future neural network oscillatory activity. Our experience of reality is associated with the global pattern of increased-frequency neural network activity.

The activity of neurotransmitters is actually produced by neuroreceptors. Neuroreceptors are complex protein structures that neurons produce and insert into their outer membrane (primarily in their dendrites). Neurotransmitters attach themselves to neuroreceptors. Neuroreceptors respond in a way that promotes or inhibits the activity of a nerve cell. Neuroreceptors are the actual agents of change in a neuron. A neuron's response to neurotransmitters is determined by the neuroreceptors it produces. Evolutionary change is reflected in the complex structure of neuroreceptors while neurotransmitters have remained relatively simple molecules.

The most widespread form of neurotransmission in the brain is referred to as rapid-acting. Rapid-acting neurotransmitters are responsible for neural oscillatory activity. They enable interactive groups of neurons to maintain stable oscillatory activity throughout life. Rapid-acting neurotransmission quickly activates or inhibits the firing of a neuron. Rapid-acting neurotransmission is classified by the neurotransmitter involved.

There are two primary neurotransmitters involved in rapid-acting neurotransmission—glutamate and GABA (gamma-aminobutyric acid). Glutamate is the primary *activating* rapid-acting neurotransmitter. Glutamate promotes the firing of a neuron. GABA is the primary *inhibiting* rapid-acting neurotransmitter. GABA is produced by inhibitory neurons.

Rapid-acting neuroreceptors are referred to as ion channel neuroreceptors. In response to a neurotransmitter, ion channel receptors open channels though which ions (positively or negatively charged atoms or molecules) are able to pass in or out of the nerve cell. In the watery environment of the body, atoms or molecules tend to form charged ions. An ion may be positively or negatively charged depending upon the balance of positively charged particles (protons) and negatively charged particles (electrons) that remain associated with an atom or molecule in a watery environment. In a watery environment, salt (sodium chloride), for example, will separate into positively charged sodium ions and negatively charged chloride ions.

Rapid-acting neurotransmission is the predominant form of neurotransmission in the cerebral cortex (the outer layer of the brain that is responsible for information processing). The primate brain (including the human brain) has a six-layer cerebral cortex—a late product of evolution and referred to as the neocortex (we'll review the neocortex in more detail later). Rapid-acting neurotransmission is responsible for maintaining cortical neural network oscillation.

Generally, rapid-acting neurotransmission does not modify the future firing threshold of a neuron. By leaving the future firing threshold of neurons unchanged, rapid-acting neurotransmission enables the brain's neural network to sustain a stable frequency of oscillation. If rapid-acting neurotransmission modified the firing threshold of the neurons involved in oscillatory activity (made them more or less likely to fire in the future), the oscillatory rate would either spin out of control or come to a halt over time.

The other primary type of neurotransmission is referred to as slower-acting. Slower-acting neuroreceptors operate by releasing a molecule (referred to as a second messenger) within the nerve cell. Slower-acting neuroreceptors are much more complex than the ion channel (rapid-acting neuroreceptors). Slower-acting neuroreceptors are the product of more recent neural evolution. Unlike the rapid-acting neuroreceptors, slower-acting neuroreceptors don't permit passage of ions through the nerve cell membrane.

The neurons that produce slower-acting neurotransmitters lie below the cortex and are referred to as *subcortical* neurons. The axons of subcortical neurons extend to the lower three layers of the six-layer neocortex. Reciprocal connections form between the subcortical neurons and the lower layers of the neocortex. Neural oscillatory activity between the subcortical neurons and the lower layers of the neocortex modifies the frequency of baseline cortical oscillation in response to drives and emotional responses.

Second-messenger molecules released by slower-acting neuroreceptors are able to influence a neuron's protein production and modify the neuron's future firing threshold. Modification of a neuron's firing threshold is the basis for memory—the likelihood of firing is influenced by past experience. The interaction of the lower layers of the neocortex and subcortical neurons (cortical-subcortical interaction) is responsible for the emotional aspects of memory. The emotional aspects of memory include our drive to seek the positive and our aversion to the negative.

There is a notable exception to the rule that rapid-acting neurotransmission does not modify future neuron-firing thresholds. The exception involves *NMDA neuroreceptors.* NMDA (shorthand for N-methyl D-aspartate) receptors are rapid-acting neuroreceptors. NMDA receptors are located in the outer cortical layers that are responsible for interaction between different cortical areas—cortical-cortical interaction. NMDA-influenced cortical-cortical interaction is responsible for the more cognitive aspects of memory.

NMDA neuroreceptors are similar to other stimulatory rapid-acting neuroreceptors, with a number of important differences. As mentioned above, the primary stimulatory rapid-acting neurotransmitter is glutamate. NMDA neuroreceptors tend to be located on cortical neurons that utilize glutamate neuroreceptors. Both glutamate and NMDA neuroreceptors utilize an ion channel that permits positively charged sodium ions to enter the neuron. However, NMDA neuroreceptors also permit calcium ions to enter the neuron. Calcium ions are able to function as a second messenger (as utilized by slower-acting neuroreceptors) and influence a neuron's future firing threshold.

NMDA neuroreceptors are also voltage-dependent. Their ion channel contains a magnesium ion that blocks the free flow of ions. The magnesium ion pops out only when the neuron membrane is depolarized. Actual depolarization of the neuron depends upon the number of glutamate neuroreceptors activated. NMDA neuroreceptors don't permit ion flow until the neuron is actually depolarized.[3] NMDA neuroreceptors reinforce the activity of neurons that are actually firing.

NMDA neuroreceptors are responsible for the cortical-cortical aspects of memory. This type of memory is referred to as *declarative* memory because it includes the facts and events that we can consciously recall and recount to others. We'll talk in more detail regarding the role of NMDA neuroreceptors when we discuss memory.

So far we've learned that baseline brain function is a combination of high-speed electrical transmission with gaps (synapses) between neurons spanned by neurotransmitters. The brain's electrical activity is oscillating due to the repetitive firing of neuron loops that involve reciprocally connected areas of the brain. Reciprocal connections are characteristic of the brain.

Neural oscillation enables the brain to function as a dynamic network in which separate areas of the brain are interconnected and continually interactive.

The pattern of neural network oscillation is modified (remembered) by the action of slower-acting and NMDA neurotransmission. Subcortical neurons modify cortical neural oscillation in response to our drives and emotions. NMDA neurotransmission influences cortical-cortical neural oscillation, enabling us to remember conscious experience from day to day. Consciousness involves a global (cortical-cortical) neural network oscillation that integrates the activity of the major cortical areas.

Brain function involves connecting the dots. By connecting the dots, I mean that the brain increases the tendency of groups of neurons to fire together. The pattern of neuron firing creates our experience. We're constantly exposed to sights, sounds, and bodily experience. The brain produces a pattern of neural network interaction that brings together these sensory experiences.

Connecting the dots is a dynamic process. Neural network oscillatory activity is always present in the living brain. Even when our eyes are closed, the area of the brain that processes vision remains active. We refer to the frequency of *the brain's baseline activity as alpha waves.* When we open our eyes, the frequency of brain waves in the visual processing area instantly increases (becomes more rapid). We call these beta waves.

Beta waves occur locally in the area of the brain that is associated with the initial processing of sensory stimuli (visual stimuli when we open our eyes). The brain's sensory lobes (the occipital, temporal, and parietal lobes) have local areas for the initial processing of the sensory information we rely on most—vision, hearing, and somatic (bodily) sensation. The occipital lobe processes vision. The temporal lobe processes hearing. The parietal lobe processes somatic (bodily) sensation. We don't have specific cortical areas devoted to taste and smell.

In order for the stimulation to be consciously perceived, it must be incorporated into a more global (cortical-cortical) pattern of neural network oscillation. When this occurs, the frequency of brain waves increases (becomes more rapid) producing gamma waves. *Gamma waves reflect the integrated activity of the major cortical areas (sensory and motor).* Alpha, beta, and gamma are simply the first three letters of the Greek alphabet.

The brain connects the dots by increasing the oscillatory frequency of those portions of the cortical-cortical neural network that are actively contributing to our experience of reality. Reality is defined by the pattern of gamma wave oscillatory activity.

The frequency of neural network oscillation is increased by reducing the baseline cortical inhibition. Cortical inhibition is the product of the rapid-acting neurotransmission involving GABA. The layer IV of the lower cortex and layer II of the upper cortex are predominantly populated with neurons that produce the inhibiting rapid-acting neurotransmitter GABA. These neurons in cortical layers II and IV are referred to as interneurons. They interact with activating rapid-acting neurons (those utilizing glutamate) in layers II, III, V, and VI.

GABA-producing interneurons are responsible for a baseline level of cortical inhibition in the upper and lower cortical layers. A localized decrease in lower cortical layer inhibition is associated with beta wave activity. Decrease in upper-level cortical-cortical inhibition (associated with focused concentration) is associated with gamma wave activity. Modification of cortical inhibition permits virtually instantaneous modification of cortical oscillatory frequency in response to sensory stimuli and focus of conscious attention. This moment-to-moment process enables us to direct behavior rapidly and efficiently.

In addition to moment-to-moment gamma wave activity, I'm able to remember some aspects of my experience on the following day. Memory requires a more enduring mechanism for connecting the dots. Memory involves a strengthening of the likelihood that neurons that fired concurrently in the past will do so in the future. Second-messenger slower-acting neurotransmission plays a prominent role in forming emotional memories. Emotional memories involve the interaction of subcortical nuclei with the lower layers of the cerebral cortex.

NMDA neurotransmission, on the other hand, is required to form memories of a more cortical-cortical nature—declarative memories. Declarative memory refers to our ability to recall, organize, and express past experience. Declarative memory requires cortical-cortical interaction of different areas of the brain through the outer cortical layers.

"Reality" refers to the patterns in which the dots are connected— the pattern of gamma wave neural network oscillation. I put quotation marks around the word "reality" to emphasize that it is the product of brain activity. Each of us experiences our own reality. Once connections have been strengthened, we are biased toward experiencing those features together in the future. *Reality is shaped by past experience.*

Emotional reactions have a particularly strong influence upon the patterns in which our brains organize information. Our emotions impact brain function from early infancy. Our emotional reactions and emotional memory are functional well before the maturation of cortical-cortical processing. They provide a foundation upon which our adult (cortical-cortical) perception of reality is constructed.

While past emotional reactions provide a foundation for cortical-cortical function, my current emotional reactions influence which of today's events I'll remember tomorrow. I'll remember any seriously negative experiences I encounter. I'll also remember enjoyable events that I'd like to repeat in the future. Throughout life, we behave in a fashion that promotes positive feelings and minimizes negative experiences.

The brain remembers patterns of neural network activity by lowering the future firing threshold between involved neurons. Lowering the firing threshold between neurons increases the likelihood that they will fire together in the future. Slower-acting neurotransmitters (produced by subcortical neurons that influence lower-layer cortical neurons) enable emotional factors to play a prominent role from earliest infancy. Even an infant seeks to repeat pleasurable experiences and avoid painful ones. NMDA neurotransmitters (associated with the upper-layer cortical neurons) permit neural firing threshold modification between neurons from different cortical areas.

Current perception is strongly biased by past perception. We're highly biased toward facial recognition, for example. Facial recognition is practiced from early childhood. As a result, we tend to see faces even when looking at inanimate objects, such as the moon or a rock formation. Illusionists rely upon our tendency to see what we're accustomed to seeing.

Neural network oscillation is required for brain function but only begins to explain how the brain works. Our brains must transform sensory stimulation into an integrated experience of reality, and our behaviors must be responsive to the reality we experience. In order to transform sensory stimulation into experience, the pattern of neural network oscillation must be modified almost instantly and be able to change from one moment to the next. The brain's reaction to stimulation is so rapid and precise that we confidently equate the brain's reality construct with the actual world around us.

In addition, we must also be able to remember past patterns of experience, particularly those that are emotionally significant. We must also develop a sense of self and learn to behave in a fashion that promotes our survival. Our sense of self plays a pivotal role in orienting us to our surroundings and guiding our behavior. We'll have to address these issues, but we can only touch on them in this introductory section.

Let's take a moment for another quick review. Localized sensory stimulation increases the frequency of neural network activity (produces beta wave activity) in the localized sensory processing area. Localized sensory processing areas of the brain are located in the parietal (for somatic or bodily sensation), occipital (for sight), and the temporal lobes (for hearing). Somatic sensation includes touch, temperature sense, position sense, and vibration sense.

For somatic sensation, sight, and hearing to be brought together into our conscious experience of reality, a further, more global increase in oscillation frequency is required that involves both the motor and sensory lobes of the brain. These oscillatory patterns can be revisited (remembered) due to modification in the firing threshold of the neurons involved. Modification of neuron-firing thresholds in the past biases our perception of the world we see around us now.

Brain function involves extensive parallel processing. Parallel processing refers to the performance of multiple operations simultaneously. The individual components of vision (color, shape, motion) are processed simultaneously with individual components of hearing (intensity, pitch, tone) and somatic sensation (touch, vibration, temperature sense, position sense). Vision, sound, and bodily sensations are brought together almost instantly into an integrated experience of reality in the multisensory processing area. The multisensory processing area lies in the parietal lobes (right and left) between the localized areas that specialize in processing sight, sound, and somatic sensations.

The brain's parallel processing activity is somewhat like an orchestral symphony. The orchestral instruments, all playing at the same time, contribute to the symphonic product. Light waves, sound waves, and somatic sensations are simultaneously translated into the currency of the brain—a frequency change in oscillating neural network activity. Once translated into the dynamic medium (frequency) of neural network activity, they are integrated into our experience of reality.

The brain's bidirectional oscillatory neural network activity makes parallel processing possible. Sensory information from the eyes, ears, and body is received by different areas in the brain's sensory gateway, the *thalamus*. The thalamus is connected by reciprocal oscillatory activity to the primary sensory processing areas in the brain (the visual, auditory, and somatic sensation processing areas). These specialized sensory areas are connected by oscillatory activity with the brain's multisensory association area.

The multisensory association area integrates sight, sound, and somatic sensation. The multisensory association area, in turn, is connected by oscillatory activity to the forward areas of the brain that are responsible for behavior. More specifically, the multisensory association area is connected by oscillatory activity to the most forward area of the frontal lobe, referred to as the *prefrontal cortex*. The prefrontal cortex is responsible for our sense of self and for guiding motor behavior, which is produced by neurons in the more rearward portions of the frontal lobe. I'll discuss all of these areas of the brain in part I, which reviews basic brain anatomy, following this introduction.

The brain's oscillatory neural network is a dynamic medium in which the oscillatory activity in one area of the brain can influence and be influenced by oscillatory activity in other areas. It permits virtually instantaneous interaction between different areas of the brain. Due to the bidirectional nature of brain oscillatory activity, the prefrontal cortex is able to receive sensory information, to focus sensory perception, and to guide behavior at the same time.

Due to continual oscillatory neural network activity, the brain has an extremely high metabolic rate. Although the brain represents only about 2 percent of our body weight, it consumes over 16 percent of the body's energy. By way of contrast, our skeletal muscles (the muscles responsible for moving the body) make up approximately 40 percent of our body weight. These are the muscles that enable us to walk, run, breathe, blink, move our eyes, talk, and eat. Despite their much-greater mass and their involvement in overall physical activity, they consume less than 15 percent of the body's energy.

The brain's pound-for-pound metabolic activity is much higher by far. Due to its high metabolic rate, the brain has increased vulnerability to oxygen deprivation. Oxygen deprivation, as sometimes experienced by drowning victims, may permanently damage the brain while leaving the skeletal muscles intact.

Repetitive patterns of activity, of which oscillation is an example, underlie all complex systems. Complex systems are identified by their pattern of activity. Repetition is essential for sustained activity. Our solar system, for example, consists of planets in repetitive orbit around the sun. Our circulatory system involves blood cells in constant circulation from the heart to the body and back to the heart. Likewise, the prevailing patterns of atmospheric and oceanic circulation are repetitive patterns of activity associated with solar energy and the earth's rotation.

Hopefully, this introductory overview leaves you eager for me to make my case in more detail and to clear up any terms that you found confusing. I'll start by reviewing basic brain anatomy in part I. I've divided this section into microscopic and macroscopic neuroanatomy. Microscopic neuroanatomy refers to the features of the brain that are not visible to the naked eye—nerve cells and the key molecules (neurotransmitters and neuroreceptors) that they rely on to function. Macroscopic neuroanatomy refers to the areas of the brain that are visible to the naked eye—white matter, gray matter, and the various anatomical parts of the brain, such as the lobes of the cerebral hemispheres, the cerebellum, and the brain stem.

Part II is devoted to the biological aspects of brain development and function. In this section, I'll present a more detailed explanation of brain function, building upon our review of brain anatomy. I'll explain that brain function is an emergent phenomenon based upon oscillatory neural network activity. I'll also explain that the brain, along with every other complex system, is subject to abnormal patterns of activity referred to as system discontinuities. "Emergent phenomena" and "system discontinuities" are terms associated with complex systems and will be addressed as such in part IV.

In part III, I'll discuss how brain function shapes our experience of the world (our psychological perspective) and our relationships with one another (our social interaction)—the psychosocial aspects of brain function. Our sense of reality is strongly influenced by the brain's reward (approach)/avoidance system. We are highly biased toward a narcissistic and unrealistic (often delusional) understanding of reality. We tend to categorize reality into "good" and "evil" based upon our reactions. This tendency underlies a great deal of human conflict.

I'll also emphasize the importance of affinity in social relationships. By affinity, I'm referring to intermediate form of bonding—less strong than the tight bonds of solid matter but definitely stronger than the virtual absence of bonding associated with chaos. Every complex system, including social systems, involves an intermediate level of bonding. There are degrees of affinity. Larger social units, such as nation states, rely upon a less intense and more inclusive form of social bonding than one typically finds in a family or tribe. The tight bonding of a tribal society can be disruptive in the context of a nation state.

Part IV of the book focuses on complex systems. The principles that underlie complex systems provide a valuable source of insight into brain development and function. My confidence in proposing neural network oscillation as the key to understanding brain function reflects my growing awareness of the importance of repetitive phenomena in complex systems.

While my own background is in clinical psychiatry, this book brings together information from both neuroscience and complex-systems theory. I hope that experts in those fields won't be offended by my attempts to integrate features of their work into a general explanation of brain function. I believe that a psychiatrist with a background in medical science and a professional lifetime as a student of human behavior is in a unique position to propose an explanation of brain function that brings together psychiatric clinical experience, the growing body of neuroscience research, and the developing field of complex systems.

PART I
BASIC BRAIN ANATOMY

I've divided brain anatomy into two categories: microscopic and macroscopic. *Microscopic* refers to the cellular components of the brain—features too small to be seen with the naked eye. This section includes the molecules used by nerve cells to communicate with each other (neurotransmitters and neuroreceptors). *Macroscopic* brain anatomy refers to features that are visible to the naked eye.

A. MICROSCOPIC

CELL STRUCTURE

Cells are the basic building blocks of living organisms. The smallest life-forms consist of one cell only. Each cell is surrounded by an external, protective membrane that provides a holding environment in which the cell's life functions can be sustained. Figure 1 (created by the US Department of Energy Genomic Science program and found at the website http://genomicscience. energy.gov—permission for use not required) shows the cellular components. The term "holding environment" is often associated with complex systems and will be discussed in part IV.

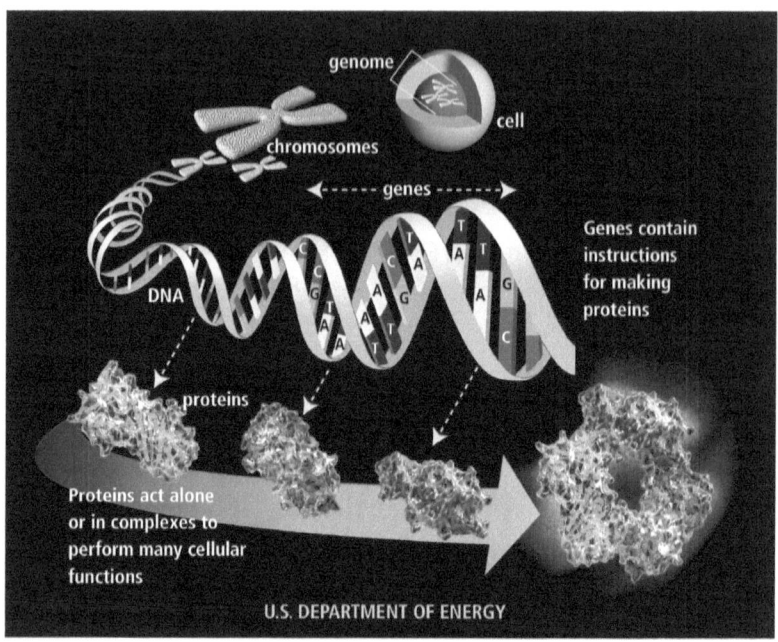

FIGURE 1: CELLULAR COMPONENTS

The nucleus is at the center of the cell (the central area of the cell in figure 1). The nucleus is a spherical structure that contains the cell's chromosomes. It is surrounded by a nuclear membrane, which separates the nucleus from the rest of the cell.

The chromosomes are long, helical strands of DNA (deoxyribonucleic acid). They're too large to pass through openings in the nuclear membrane and remain within the nucleus. Humans have forty-six chromosomes—twenty-three chromosomes from each parent. Included in the forty-six chromosomes are the X and Y chromosomes that determine sexual development. In most cases, females have two X chromosomes while males have one X and one Y chromosome.

Each chromosome contains numerous genes. Genes are simply sections of a chromosome's long DNA strand. It's estimated that our chromosomes contain roughly 20,000–25,000 genes.[4]

The outer surface membrane of the cell is referred to as a

phospholipid membrane. Phospholipids are a class of lipid molecules that make up the outer cell membrane. A cell's outer phospholipid membrane is much less permeable than the nuclear membrane. It forms a robust barrier between the inside and outside of the cell.

PROTEINS

Cell structures responsible for protein production are found between the nuclear membrane and the outer cellular membrane. Proteins are the operating tools the cell uses to sustain life and interact with the cellular environment. Proteins are chains of amino acids (relatively simple molecules referred to as the building blocks of life).

In order for proteins to be produced, the genetic information on the DNA template must be transported outside the nucleus to the structures responsible for protein production. The information on a specific gene is copied by RNA (ribonucleic acid) molecules. RNA molecules are much smaller than the chromosomes and pass easily through the nuclear membrane. Once outside the nucleus, RNA copies serve as templates for producing specific proteins.

It's important to keep in mind that proteins are three-dimensional. Notice the three-dimensional structure of the protein in figure 1. The three-dimensional configuration of a protein (how the protein is folded) can change significantly with even minor changes in the sequence of amino acids that make up a protein molecule. Variability in the three-dimensional configuration of proteins enables them to play a prominent role in biological evolution. Variations in the three-dimensional configuration of proteins that promote survival and reproduction are more likely to be transferred across generations.

The human body is built from a single cell, and every cell of the body relies upon the genetic library maintained in our chromosomes. Different cell types in the body use different portions of the gene library for protein production. A liver cell, for example, requires a different pattern of protein production than a nerve cell. The pattern of protein production by a specific cell depends upon the cell type and its stage of development.

Neuroreceptors enable nerve cells to respond to neurotransmitters released by other nerve cells. Neuroreceptors are proteins found within the outer nerve-cell membrane. They're protein islands resting in the neuron's outer phospholipid membrane.

Neuroreceptors span the neuron's outer phospholipid membrane. The outer portion of the neuroreceptor is accessible to chemicals (including neurotransmitters) outside the neuron's phospholipid membrane. Most of the neuroreceptor molecule lies within the phospholipid membrane. However, the inner portion of the neuroreceptor is exposed to the interior of the nerve cell. Neuroreceptors serve as gateways of interaction between nerve cells.

Neurotransmitters are chemicals released by a neuron. Neurotransmitters released by a nerve cell become attached to the neuroreceptors of another nerve cell. When this occurs, the neuroreceptor reacts in a manner that alters the internal environment of that nerve cell.

There are two primary types of neuroreceptors—rapid-acting and slower-acting. Rapid-acting neuroreceptors open to form an ion channel that permits ions to passively move through the membrane. Slower-acting neuroreceptors, on the other hand, release a second messenger (ion or molecule) within the nerve cell. Neurotransmitters released by one nerve cell interact with the neuroreceptors of another. In this way, nerve cells are able to stimulate or inhibit the activation of other nerve cells.

NERVE CELLS AND NEUROTRANSMISSION

Nerve cells make up the communication matrix of the brain. They are the cells that enable the brain to function as an information-processing system. Supporting cells, referred to as glial cells, provide a holding environment in which nerve cells function.

Nerve cells consist of dendrites, a cell body, and an axon. The cell body is the central portion of the nerve cell that contains the nucleus. Dendrites and axons are branch-like extensions that protrude from the cell body. Nerve cells typically have a large number of dendrites and a single axon. The axon is usually a long extension that reaches out to other nerve cells. Axon tips release neurotransmitters. Dendrites, on the other hand, are typically much shorter extensions. Dendrites contain neuroreceptors in their external membranes. Neurotransmitters released by the axon tip of one neuron interact with neuroreceptors on the dendrites of another neuron.

Nerve-cell activation usually begins in the dendrites. Neurotransmitters released by an adjacent nerve-cell axon tip attach themselves to a dendrite's neuroreceptors. The impact may stimulate or inhibit nerve-cell activation. A sufficiently high level of stimulatory neurotransmitters will activate the dendritic membrane.

Once started, activation passes from the dendrites to the cell body and then proceeds from the cell body down the axon. When neural activation reaches the axon tip, neurotransmitters are released into the narrow space (synaptic gap) separating the axon tip from the dendrite of an adjacent nerve cell. Once stimulatory neurotransmitters cross the synaptic gap and activate a sufficient number of neuroreceptors, the process is repeated. Transfer of neural activation from one neuron to another is a one-way process.

Nerve-cell activation requires an energy gradient across the outer nerve-cell membrane. "Energy gradient" is another term often associated with complex systems and will be discussed in more detail in the final section (part IV) of the book. In a resting state, nerve cells maintain a difference in the concentration of positively charged ions between the inside and outside of the neuron's outer phospholipid membrane. Positively charged sodium ions predominate outside the nerve-cell phospholipid membrane, while positively charged potassium ions predominate inside the nerve cell. In a resting state, the region outside of the nerve cell is more positively charged than the inside. The electrical potential (energy gradient) across the nerve-cell external membrane is so small that highly specialized instruments are needed to measure it.

Nerve cells maintain an energy gradient across their cell membranes by using energy to pump positively charged sodium ions from inside the cell to the outside and positively charged potassium ions outside the nerve cell to the inside. The pumping action required to maintain the necessary energy gradient across nerve-cell membranes is a primary contributor to the brain's high level of metabolic activity.

The nerve-cell membrane becomes less polarized when positively charged sodium ions enter the cell through sodium ion channel neuroreceptors in response to neurotransmitters. If sufficient sodium ions enter the cell, an area of the nerve-cell membrane is depolarized; there is no longer an energy gradient between the outside and inside of the nerve cell in that area. Depolarization in one area of the neural membrane opens the ion channels of adjacent sodium ion channel neuroreceptors. Neural activation is transmitted within a neuron through a wave of membrane depolarization that spreads from the point of origin. Once started, depolarization passes rapidly from dendrites, through the cell body, and down the nerve-cell axon.

Nerve-cell depolarization is initially ended by releasing positively charged potassium ions from inside to outside of the cell through potassium ion channels that are sensitive to voltage changes in the nerve-cell membrane (voltage-gated). The release of potassium ions from inside to outside the nerve cell immediately restores the energy gradient. However, this is only a short-term fix. Between depolarization events, the nerve cell requires energy to pump sodium ions out of the cell and potassium ions into the cell. The pumping activity that occurs between firing events is needed for a nerve cell to fully reestablish the ionic state that existed prior to depolarization. Pumping activity between depolarization events permits the neuron to fire over and over again. Repetitive firing is required for neural oscillation.

The firing of a neuron is a bit like the action of a spring-powered mousetrap. Energy is needed to pull the trapping lever into place and hold it in a ready position. The trapping lever is held in position until it's released by a gentle nudge of the bait. When released, the potential energy in the spring rapidly snaps the trapping lever shut.

In a similar manner, the energy gradient across the neuron cell wall is a form of potential energy. Energy is required to maintain the energy gradient in a state of readiness to fire. Potential energy is released by "nudging" the neuron with neurotransmitter molecules released into the synaptic cleft. Once activated, electrical activity passes rapidly through the body of a nerve cell and down the nerve cell axon.

MYELINATION INCREASES AXON SPEED

The speed of electrical transmission along an axon is significantly increased by *myelination*. One of the functions of glial cells (the cells that provide a supportive environment for nerve cells) is to form white-matter sheaths around developing axons—axon myelination. Axon myelination increases the speed and efficiency of transmission (somewhat like the introduction of an express highway improves the efficiency of a two-lane road).

Axon myelination is characteristic of the long axons that connect different areas of the brain with one another. Axons that contribute to the reciprocal connections between different areas of the brain are continually active due to neural network oscillation. Continual axonal activity associated with neural network oscillation prompts the formation of myelin sheaths. Myelin gives the axons a white appearance. Most of the brain's mass is referred to as white matter due to the presence of myelinated axons.

Myelination increases the rate of electrical transmission along the axon by permitting the depolarization process to leap from one node of Ranvier to the next. Nodes of Ranvier are small areas of exposed axon nerve membrane that lie between areas of axon myelination.[5] Speed is critical to the overall efficiency of the brain's ability to process information. The brain requires near instantaneous processing for even everyday activities, such as language fluency, walking, hitting a baseball, or catching a Frisbee.

Nerve cells pass on information from one nerve cell to the next by releasing chemicals called neurotransmitters. The axon tip of a nerve cell releases a neurotransmitter that is specific to that nerve cell. We classify nerve cells by the neurotransmitter they release. Neurotransmitters released by a nerve cell bind to neuroreceptors on an adjacent nerve cell. By binding to the neuroreceptors of another nerve cell, neurotransmitters influence the activity of the postsynaptic neuron. Typically, the action of a neurotransmitter either stimulates or inhibits the firing of the postsynaptic neuron.

RAPID-ACTING AND SLOWER-ACTING NEUROTRANSMISSION

Neurotransmission may be rapid-acting or slower-acting depending upon the type of neuroreceptor involved. Rapid-acting neurotransmission is used when transmission speed is critical. Rapid-acting neurotransmission involves opening a pore in the neuroreceptor that permits ions to passively enter or leave the neuron. Rapid-acting neurotransmission is responsible for maintaining neural oscillatory activity.

Rapid-acting neurotransmission permits neurons to interact very rapidly with one another without modifying a neuron's future firing tendencies (with the exception of NMDA neurotransmission). Slower-acting neurotransmission, on the other hand, plays more of a role in the long-term management of neural firing. Slower-acting neurotransmission allows the child to learn from positive and negative emotional experiences and modify future behavior accordingly.

Slower-acting neurotransmission utilizes a second-messenger receptor system. Instead of an ion passing through the receptor, a specialized molecule (referred to as a second messenger) is released inside the cell of the neuron. The second messenger can actually influence the cell's future activity by influencing the production of specific proteins from the cell's DNA library. By modifying the production of specific proteins, subsequent neural activity is modified by experience.

When we think of human brain function, we are actually thinking of neural network function that has already been transformed by experience. Rapid-acting neurotransmitters are the workhorses of the brain. They enable a chain of neurons to sustain oscillatory activity indefinitely. However, brain function, as you and I experience it, is the product of a neural network that has already been modified by experience beginning in early childhood. Second-messenger neurotransmitters produced by subcortical neurons interact with the inner layers of the cortex from earliest childhood. Second-messenger neurotransmitters establish an emotional foundation for future cognitive development and influence cognitive processing throughout our lives.

The outer cortical layers, those responsible for the cognitive (cortical-cortical) aspects of memory, become functional later. The outer cortical layers responsible for cortical-cortical interaction utilize NMDA neuroreceptors—a specialized form of ion channel, rapid-acting neurotransmission. In addition to sodium ions, NMDA receptors permit calcium ions to enter the nerve cell, where they act as second messengers to modify a neuron's future firing threshold. NMDA receptors will be addressed in more detail later in the book when we discuss memory.

B. MACROSCOPIC

Macroscopic neuroanatomy refers to the brain structures that are visible to the naked eye. The brain is divided into gray matter and white matter. Neuron cell bodies are located in the gray matter. Gray matter is located in either specific layers (such as the cerebral cortex) or in localized areas referred to as nuclei (subcortical nuclei). Most of the brain tissue is white matter. White matter is composed of myelinated neural axons.

Each cerebral hemisphere has an outer layer of gray matter referred to as the *cerebral cortex*. The cortex is an outer layer covering the cerebral hemispheres formed by nerve cells. Microscopically, the human (and primate) cortex contains six identifiable nerve cell layers and is referred to as *neocortex*. Directly under the cerebral cortex lies white matter containing myelinated axons that carry information to and from the cerebral cortex to other areas of the brain and body.

OUTER BRAIN

The brain is divided into the two cerebral hemispheres (right and left), the brain stem, and the cerebellum. The outer portion of each cerebral hemisphere is divided into four lobes: frontal, parietal, occipital, and temporal lobes, as indicated below in figure 2 (used with permission by Elsevier, it is figure 5.4 in the book *Cognition, Brain, and Consciousness* by Bernard J. Baars and Nicole M. Gage, published by Elsevier in 2010).

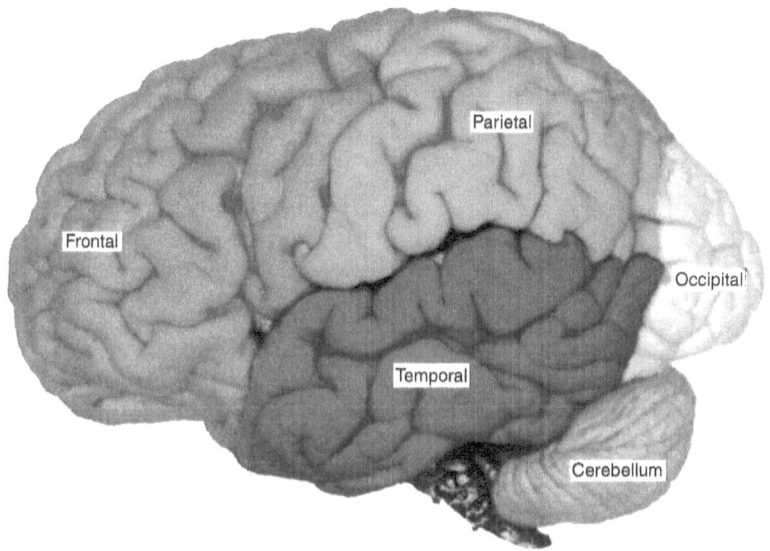

FIGURE 2: THE OUTER BRAIN

The frontal lobe lies forward, directly behind the forehead. It is directly involved in voluntary muscle movement and is often referred to as the motor lobe. The anterior frontal lobe is referred to as the prefrontal lobe. The prefrontal lobe is responsible for evaluating and directing voluntary motor movement. The posterior portion of the frontal lobe is directly responsible for voluntary muscle movement.

The parietal, occipital, and temporal lobes lie directly behind the frontal lobe. These lobes are responsible for receiving and processing the sensory input that we rely upon (the sensory lobes). The anterior area of each parietal lobe processes somatic sensation (touch, vibration, temperature sensation, pain, and position sensation). Light perception (vision) is processed by the occipital lobes, and sound perception (hearing) by the temporal lobes.

SENSORY LOBES

The primary somatic sensory cortex lies in a strip along the anterior border of each parietal lobe (adjacent to the frontal lobe). Sensory information concerning touch, vibration, temperature sensation, pain, and position sensation is initially received by the primary somatic sensory cortex. Directly behind and adjacent to the primary somatic sensory cortex is an area of cortex that processes the somatic sensory information. I'll refer to this type of sensory processing area as unimodal because it processes only somatic sensory information. The somatic unimodal area assembles the elements of somatic sensation into coherent bodily experience.

The primary visual sensory cortex lies in the posterior portion of each occipital lobes. Sensory information required for vision (such as color, linearity, brightness, and contrast) is initially received by the primary visual sensory cortex. Surrounding and adjacent to the primary somatic visual cortex is the visual unimodal processing area that is responsible for assembling the elements of visual sensation into a coherent visual experience.

The primary auditory sensory cortex lies in a portion of each temporal lobe. Sensory information required for hearing (such as pitch, rhythm, and volume) is initially received by the primary auditory sensory cortex. Surrounding and adjacent to the primary auditory cortex is the auditory unimodal processing area that is responsible for assembling the elements of auditory sensation into a coherent hearing experience.

Each parietal lobe also contains a multisensory association area. The multisensory association area lies in the posterior portion of each parietal lobe and is surrounded by the somatic, visual, and auditory unimodal processing areas. The multisensory association area is responsible for integrating the somatic, visual, and auditory information in each hemisphere.

Keep in mind that the right side of the brain deals with sensory information from the left side of the body while the left side of the brain deals with the right side of the body. In addition to bringing the components of sensation into a coherent experience, the right and left multisensory association areas interact with one another, enabling the brain to provide a coherent overall experience of reality from one moment to the next.

FRONTAL (MOTOR) LOBE

The frontal lobe processes motor activity. It's divided into three functional areas that are analogous to areas in the sensory cortex. The primary motor cortex is analogous to the primary somatic cortex. The supplementary motor area is analogous to the unimodal somatic sensory processing area. The prefrontal cortex is analogous to the multisensory association area. The prefrontal cortex is often referred to as the anterior association area while the multisensory association area is referred to as the posterior association area.

The primary motor cortex lies along a strip in the posterior portion of each frontal lobe (adjacent to the border of the parietal lobe and parallel to the primary somatic cortex). The axons of the primary motor cortex directly impact muscle activity in a fashion analogous to the initial cortical activation of the primary somatic cortex by somatic sensory stimulation. Neurons in the primary motor cortex send their axons to neurons in the spinal cord. Neurons in the spinal cord, in turn, send axons directly to voluntary muscles. The right side of the brain controls motor activity on the left side of the body and vice versa.

The supplementary motor cortex lies directly in front of the primary motor cortex. The supplementary motor cortex enables the primary motor cortex to smoothly engage in coordinated motor activity with minimal conscious effort. The supplementary motor cortex translates our intended action into coordinated motor activity (analogous to the unimodal sensory somatic processing area which takes the elements of somatic sensation and brings them together into an integrated somatic experience). Even a relatively simple activity, such as walking, requires a complex pattern of overlearned motor activities. An early toddler doesn't find walking to be even remotely second nature; conscious effort is required for each step, with frequent missteps. Adults, however, have been walking since childhood and the activity has been repeated so often that it is automatic.

The supplementary motor cortex is critical for speech—a form of overlearned activity that we humans rely upon heavily. An area of the left supplementary motor cortex is essential for speech production. Injury to this area (referred to as the cortical speech area) renders individuals unable to produce intelligible speech despite knowing what they want to say. A normally functioning left cortical speech area permits me to focus upon the ideas I want to express without having to worry about fluently producing the words and phrases needed to express my views.

The prefrontal cortex directs our activities. It serves an integrating function analogous to the multisensory association area. Reciprocal connections with the multisensory association area provide the prefrontal cortex with moment-to-moment information concerning our somatic experience and visual/auditory events around us. The prefrontal cortex integrates our sensory perception with our striving for the positive while avoiding the negative. The prefrontal cortex enables us to maintain a coherent sense of self and to behave in a fashion that promotes our self-interest.

BRAIN STRUCTURE

In figure 2, we only see the cortical surface. The following figure 3 (used with permission by Elsevier, it is figure 5.10 in the book

Cognition, Brain, and Consciousness by Bernard J. Baars and Nicole M. Gage, published by Elsevier in 2010) permits us to look at the inner structure of the brain in a stepwise fashion. At the center of the brain, hidden by the outer cerebral hemispheres, are the thalami (plural of thalamus—a bilateral structure present in both the right and left hemisphere). Below the thalami lies the brain stem. The brain stem connects the cerebral hemispheres with the spinal cord. Directly behind the brain stem and below the cerebral hemispheres lies the cerebellum.

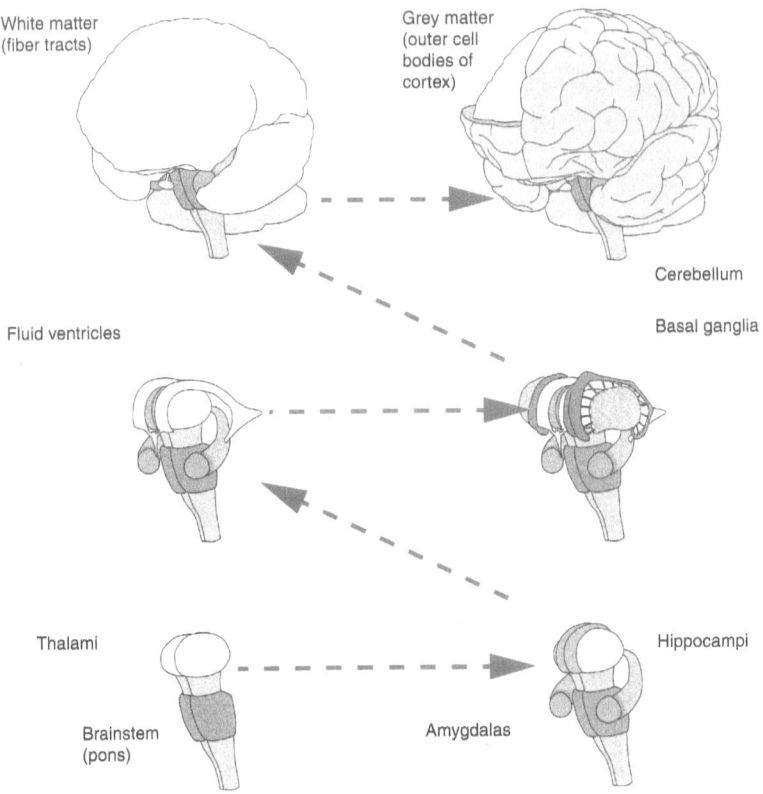

FIGURE 3: INSIDE THE BRAIN

BRAIN STEM

The brain stem reflects the earliest phases of brain evolution. The brain stem controls a wide range of basic life-support activities that are handled automatically without our having to consciously think about them, such as heart rate, respiratory rate, and hunger.

CEREBELLUM

The cerebellum lies under the cerebral cortex and behind the upper brain stem. The cerebellum has a three-layer cortex that reflects an intermediate phase of evolution between the brain stem and the six-layer neocortex of the cerebral hemispheres. The cerebellar cortex has a mixture of inhibitory and excitatory cellular layers. Voluntary muscles don't function in isolation. They tend to have antagonistic relationships in which contraction of one muscle group must be accompanied by relaxation of another muscle group. In order to flex your biceps muscle, for example, your triceps muscle must relax.

In contrast to the supplementary motor cortex and basal ganglia (subcortical nuclei involved in movement that are discussed in the following section), the cerebellum ensures that the movement is smooth by coordinating the action of opposing muscle groups. Smooth movement requires the contraction of one group of voluntary muscles and relaxation of opposing muscle groups. Cerebellar abnormality is associated with tremor due to difficulty coordinating flexion and extension of opposing muscle groups. The cerebellum coordinates the inhibitory and excitatory activity needed for voluntary movement.

INNER CEREBRAL HEMISPHERE—THALAMUS AND LIMBIC AREA

The inner portion of the right and left cerebral hemisphere contains the right and left thalamus. The right and left thalamus are the gateway to the brain of sight, hearing, and somatic sensation (touch, temperature sense, vibration sense, and position sense). Sight, hearing, and somatic sensation arrive at the thalamus before going to the primary sensory areas of the cortex. These are the senses that we humans rely upon most heavily. They enable us to react to sight, sound, and bodily sensation. Seeing and hearing are particularly useful because they permit us to detect activity at a distance. Somatic sensation provides immediate information regarding the status of our bodies. We also have a sense of smell and taste but do not have specific neocortical sites devoted to them.

There are strong reciprocal connections between the thalamus and the sensory cortex. The vast majority of fibers received by the thalamus are from the cortex rather than from sensory organs. Reciprocal connections between the primary sensory areas and the thalamus are responsible for oscillatory activity between these areas.

I refer to the area surrounding the thalamus as the limbic area. The term limbic refers to border or margin. The limbic area of the right and left hemispheres contains a number of structures seen in figure 3 above, including the hippocampus, amygdala, and basal ganglia. These neural cell clusters are referred to as nuclei.

The nuclei in the limbic area and the brain stem are referred to as subcortical nuclei because they are found below the outer layer of the brain referred to as the cortex. The hippocampus is needed for declarative memory. The amygdala plays a prominent role in our emotions and avoidance system. The basal ganglia process voluntary movement. The nucleus accumbens (included in the basal ganglia) plays a prominent role in our dopamine reward system.

The areas of the frontal lobe, particularly the supplementary motor area, are interactive with the basal ganglia (see figure 3). The basal ganglia play a prominent role in the actual performance of the motor area activity. They contain a complex combination of stimulatory and inhibitory pathways needed for producing overlearned movements, such as walking and talking. Abnormalities involving the basal ganglia are associated with abnormal slowing or bouts of uncontrolled motor activity.

Drives and emotions shape our perception of reality—our sense of good and evil, our sense of self, and our relationships with others. The nucleus accumbens and the amygdala are key contributors to our approach/avoidance reactions, which will be discussed further in part III of this book. They develop reciprocal interaction with the lower layers of the neocortex in earliest infancy.

As we'll discuss in detail as the book progresses, the cortical layers develop in a particular order, starting from the innermost (lowest), followed by the upper layers. The lower layers are the first to become fully functional. Those layers are responsible for cortical-subcortical interaction. The lower layers of the neocortex provide a foundation for the development of the upper layers. The upper layers (the last to become fully functional) are responsible for cortical-cortical interaction.

Moment-to-moment experience involves both the upper and lower layers of the cerebral cortex. The lower layers are involved in emotional processing and emotional memory. The upper layers are responsible for information processing—processing information on a moment-to-moment basis. Information memory (referred to as declarative memory, reflecting our ability to report past events) relies upon NMDA neuroreceptors that reinforce the firing of activated neurons in the upper layers of the cerebral cortex.

Information memory (declarative memory) may be short-term (a matter of minutes) or long-term (days later, for example). Short-term declarative memory relies upon NMDA neuroreceptors in the upper cortical layers (cortical layers II and III). These neuroreceptors reinforce the firing of activated upper-cortical-layer neurons.

Long-term declarative memory requires further reinforcement by the hippocampus. The hippocampus is a prominent subcortical location of NMDA neuroreceptors. Without hippocampal involvement, we are able to remember events for a brief period only. With hippocampal involvement, we are able to recall events days, weeks, and years later. Long-term cortical-cortical memory is extremely important. Long-term memory is required to construct the explanations (narratives) that we use to gain understanding of our experience (later I'll discuss the term "narratives" in more detail).

The bilateral basal ganglia (which include the right and left nucleus accumbens) process movement. I can scratch my head with very little conscious awareness of the motion involved. The basal ganglia enable me to perform such movements in an automatic fashion. This is in sharp contrast to a young infant's behavior. The motions of an infant's arms and legs are poorly coordinated. They tend to be jerky and spasmodic. Through a process of endless trial and error, an infant gradually develops the increasingly more coordinated motor activity of an adult. The movements become overlearned and automatic—like walking and talking in an adult.

OUTER CEREBRAL HEMISPHERE—WHITE MATTER AND CORTEX

Surrounding the limbic area of each cerebral hemisphere is the outer cerebral hemisphere. The outer cerebral hemisphere includes white matter (fiber tracts) with an outer covering (layer) of gray matter (the cortex). The white matter consists of myelinated neural axons that connect the gray matter areas of the brain with one another and with other areas of the brain/

body. The surface covering is referred to as the cerebral cortex. In primates (including humans) the cerebral cortex has six layers and is referred to as the neocortex. The outer cerebral cortex is divided into the frontal lobe, parietal lobe, occipital lobe, and temporal lobe, as shown in figure 2 above.[6]

The six-layer arrangement of nerve cells in the human cerebral cortex is illustrated below in figure 4 (used with permission by Elsevier, it is figure 5.9 in the book *Cognition, Brain, and Consciousness* by Bernard J. Baars and Nicole M. Gage, published by Elsevier in 2010). The cortical layers develop from the inside out, starting with layer VI. The six layers of the cortex are arranged in vertical columns somewhat like the hexagonal cells of a honeycomb. Think of an individual column as a vertical cylinder with six cellular layers.

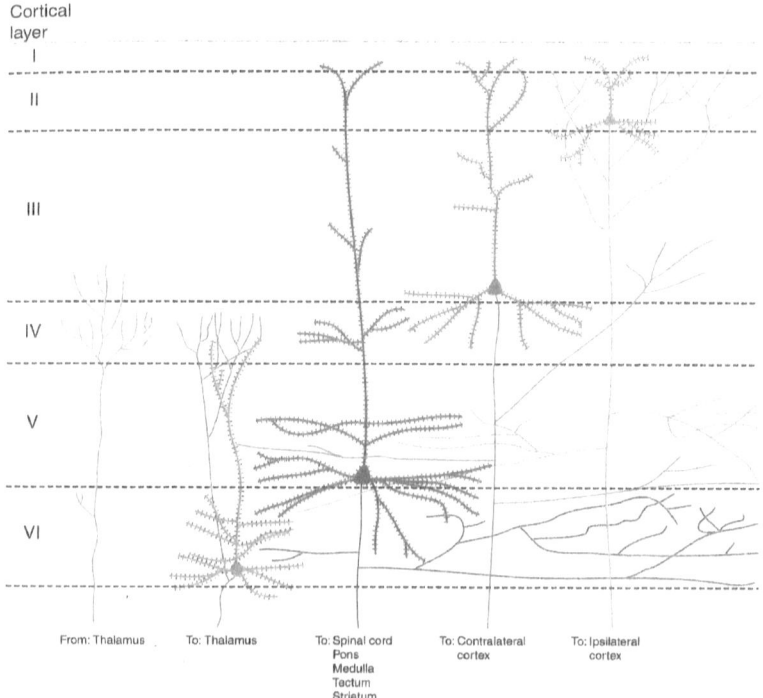

FIGURE 4: SIX-LAYER CEREBRAL CORTEX

The neurons in all six cellular layers utilize rapid-acting neurotransmitters to rapidly act on other nerve cells. In general, rapid-action neurotransmission has no long-term consequences. Rapid-acting neurotransmitters can be activating (increase the likelihood of another neuron firing) or inhibiting (decrease the likelihood of another neuron firing).

Layers III (and to a lesser extent layer II), V, and VI utilize *activating* rapid-acting neurotransmitters. They have large pyramidal cells whose *axons reach areas outside of the cortical column*. The pyramidal cells utilize the activating rapid-acting neurotransmitter, glutamine. Layers V and VI maintain the neural oscillatory activity between the cortex and subcortical areas (the thalamus and other subcortical nuclei). The pyramidal cells of layers II and III maintain neural oscillatory activity between different cortical areas. Layer VI connects with the thalamus. It is the first of the layers to become functional and the only one that is fully functional at birth.

After birth, experience shapes the oscillatory activity of the neural network. Layers IV and V, the first cortical layers to reach functionality after birth, are engaged in oscillatory activity with the subcortical nuclei of the approach/avoidance system. The subcortical nuclei of the approach/avoidance system utilize slower-acting, second-messenger neurotransmission that shapes brain function based on our positive and negative experiences.

Layer V connects with nuclei below the cortex (referred to as subcortical nuclei). Subcortical nuclei, located primarily in the limbic area, are responsible for drives and emotions. Layer V does not become fully functional until after birth. Layer III (and to a lesser extent layer II) connects to other areas of the cortex. The cortical-cortical connections of layers II and III are the last to become fully functional.

Layers II and III provide connectivity between cortical areas. Layer II receives input from other cortical areas. The pyramidal cells of layers II and III provide output to other cortical areas.

Initially, only adjacent areas of the cortex are reciprocally connected. More distant cortical connections are completed later. Cortical-cortical memory relies upon NMDA receptors. Cortical-cortical memory is shaped by our emotional memories (both positive and negative) beginning in earliest childhood and continuing throughout life.

Layers II and IV utilize the *inhibitory* rapid-activating neurotransmitters referred to as interneurons. Layers II and IV receive information from outside their cortical column, but their *inhibitory axons operate exclusively within the cortical column.* Layer IV receives information from subcortical areas (e.g., from the thalamus and limbic area nuclei). Layer II receives information from other cortical areas.

The overall inhibitory activity within the cortical columns is responsible for a baseline frequency of neural oscillation (alpha wave activity). Incoming stimulation into layer IV from subcortical areas (thalamus and subcortical nuclei) decreases cortical column inhibition and increases the cortical oscillation rate resulting in beta wave activity. Incoming stimulation of layer II from other cortical areas (particularly the prefrontal cortex) further decreases cortical column inhibition resulting in gamma wave activity. Gamma wave activity involves cortical-cortical interaction and is more global in nature.

Most of the interneurons in layers II and IV utilize the inhibitory rapid-activating neurotransmitter, gamma-aminobutyric acid (GABA). As the cortical layers mature, the interneurons within each column are organized into layers II and IV of the mature cortex. Their axons are relatively short since they do not extend outside the cortical column.

Layer I is referred to as the molecular layer. Layer I's function is not as well understood as layers II through VI.

PREFRONTAL CORTEX AND THE MULTISENSORY ASSOCIATION AREA

A large area of the parietal lobe serves as a multisensory association area. The multisensory association area is located between the cortical areas that process specific sensory modalities (somatic sensation, vision, and hearing)—the unimodal sensory processing areas. Somatic sensation, vision, and hearing are brought together into an integrated experience of reality by the multisensory association area.

The multisensory association area in the parietal lobe is directly related to the prefrontal cortex of the frontal lobe. Reciprocal oscillatory interaction permits the multisensory association area to directly inform the prefrontal cortex of sensory experience. Reciprocal oscillatory interaction also permits the prefrontal cortex to focus our sensory experiences while guiding our motor activities. The right and left multisensory association areas bring together our sensory experiences into a coordinated experience of reality.

The prefrontal lobe is responsible for overseeing conscious brain function. The ability to delay, weigh options, and implement our best responses is extremely important for individual and social survival. These skills depend upon the controlling influence of the prefrontal lobe over emotionally driven reactions driven by limbic area structures. The long axonal connections between the prefrontal lobe and the multisensory association area are the last to reach full maturity (i.e., to become fully myelinated). Frontal lobe oversight is a work in progress during adolescence.

The frontal lobe is the only cortical lobe directly influenced by the brain's dopamine reward system. As such, the frontal lobe, through the influence of the prefrontal lobe, is able to progressively gain the degree of cortical dominance needed for adult function. The gradual strengthening of prefrontal lobe control shifts our activity away from short-term, pleasure-based toward more long-term, reality-based behavior.

PART II ————————————

BIOLOGICAL ASPECTS OF BRAIN DEVELOPMENT AND FUNCTION

The cerebral cortex consists of a well-defined layer of gray matter that covers the surface of both cerebral hemispheres. In human beings, approximately 90 percent of the cerebral cortex is divided into six layers and is referred to as the neocortex. The key to human brain function lies in cortical function.

In order to understand cortical function, we'll need to look closely at cortical development and the functions of nerve cells in the six cortical layers. Cortical development is not a straightforward linear process. Initially, there is only one cortical layer that is referred to as the ventricular zone. The ventricular zone consists of a layer of cells from which neurons and glia cells develop.

External to the ventricular zone, the first neurons to be generated establish a zone called the cortical plate. Between the cortical plate and an outer layer referred to as the marginal zone, cortical layers II through VI are formed.[7] The outer marginal zone becomes cortical layer I. The ventricular zone and cortical plate are no longer present in the six-layer cortex. Under the six-layer cortex, white matter lies directly under layer VI.

I've included this level of detail to help you understand why

I'll be focusing primarily on the development and function of layers II through VI. The ventricular zone and cortical plate are transitory and no longer present in the adult brain. The outer marginal zone becomes cortical layer I.

A. CORTICAL LAYER DEVELOPMENT

The innermost neocortical layers (IV, V, and VI) interact with subcortical neural structures while the upper layers (II and III) interact with other cortical neural structures. Layer I is not as well understood. Cortical-subcortical interaction provides us with sensory information (via the thalamus) and approach/avoidance information (via limbic area subcortical nuclei). Cortical-cortical interaction allows the brain to function as an integrated unit under the guidance of the prefrontal lobe. We'll discuss each layer in the order in which they become fully functional.

INNERMOST CORTICAL LAYER (LAYER VI)

Layer VI is the only cortical layer that is fully functional at the time of birth. Layer VI cortical function is the product of reciprocal neural connections between the thalamus and layer VI of the cerebral cortex. These connections are genetically directed. Thalamic axons and layer VI cortical axons meet and follow their opposite number to its site of origin, forming reciprocal connections that permit oscillator activity between the thalamus and layer VI of the cerebral cortex. Reciprocal connections are the basis for bidirectional electrical interactivity between areas of the brain. Bidirectional interactivity (oscillation) transforms the brain into a dynamic network of neural activity.

The oscillatory activity of reciprocal connections is strengthened by *neurotrophins*. Neurotrophins, also known as neurotrophic proteins, are proteins produced by nerve cells.[8] They are released by firing neurons and foster the development of neurons that fire simultaneously.[9] Before birth, they play a particularly important role in brain development. After birth, neurotrophins are joined by slower-acting second-messenger neurotransmission and later by NMDA neurotransmission, enabling us to develop and maintain experience-based neural network patterns of interaction.

Oscillatory neural activity fosters the preservation of neurons that are repetitively firing. Oscillation also plays a role in determining which neural axons are myelinated. The high level of neural axon activity associated with neural oscillation fosters myelination in the axons involved.

Without neurotrophic proteins, nerve cells will die off—a process that is referred to as apoptosis. Active neurons are preserved by neurotrophic proteins. Inactive neurons, on the other hand, undergo apoptosis.

The oscillatory neural network is fine-tuned by apoptosis— elimination of axons not involved in oscillatory activity. Fine-tuning is necessary to ensure robust and highly reliable network interaction between the thalamus and the cortex. In part IV of this book, when we discuss complex systems, I'll be referring to such highly organized networks of interaction as strong (more highly predictable) attractor systems. Complex systems vary in terms of their degree of organization and predictability. Brain activity requires an extremely high degree of predictability.

COMPLETION OF CORTICAL-SUBCORTICAL CONNECTIVITY (LAYERS IV/V)

Following birth, the next developmental phase of reciprocal connections involves cortical layers IV and V and subcortical nuclei (other than the thalamus). I'll focus on subcortical nuclei located in the limbic area that are involved in approach/avoidance and memory. Approach/avoidance is our tendency to seek out what we find pleasurable and avoid things we find painful. Memory enables us to use past experience (particularly those that are positive and negative) to guide future behavior. We will discuss memory and approach/avoidance further in part III.

Layer IV takes on the role of receiving incoming axons from both the thalamus and the other subcortical nuclei. The interneurons of layer IV are inhibitory and establish an inhibitory field that modulates the oscillatory frequency of cortical-subcortical interaction. Baseline cortical-subcortical oscillatory interaction (alpha wave activity) is increased (transformed into beta wave activity) locally by decreasing the inhibitory field in layer IV in response to sensory stimulation.

The brain's approach/avoidance system is the product of the nucleus accumbens and the amygdala. The nucleus accumbens is a key contributor to the brain's dopamine reward system. The amygdala is very responsive to negative stimuli. It motivates us to avoid repeating painful experiences.

The cortical-subcortical interaction of the nucleus accumbens and the brain's dopamine reward system is limited to the brain's frontal lobe (particularly the prefrontal lobe). Driven by the brain's dopamine reward system, the prefrontal lobe strives to maintain overall control of conscious behavior. The mature prefrontal lobe is able to resist short-term gain (immediate pleasure) in favor of a long-term sense of satisfaction.

The subcortical nuclei utilize slower-acting second-messenger neurotransmission. In combination with neurotrophins, second-messenger neurotransmission permits the drives and emotions to influence brain development from early infancy.

CORTICAL-CORTICAL CONNECTIVITY (LAYERS II/III)

The final phase of cortical development involves the formation of reciprocal interaction between separate cortical areas—cortical-cortical interaction. Reciprocal interaction is initially functional between cortical areas that are immediately adjacent to one another. Over time, reciprocal interaction becomes fully functional between cortical areas that are more geographically separated.

Cortical-cortical connections involve cortical layers II and III. Cortical layer II interneurons receive input from other cortical areas while the pyramidal cells of layers II and III send axons to other cortical areas. Cortical-cortical reciprocal connections form an integrated cortical neural oscillatory network. The interneurons of layer II are inhibitory and establish an inhibitory field that modulates the oscillatory frequency of cortical-cortical interaction. By decreasing the inhibitory field of layer II, the neural oscillation rate in the cortical-cortical neural network is increased in frequency to gamma wave activity. Cortical inhibition (involving cortical layers II and III) is more pronounced because it involves inhibition at two separate cortical locations—both of the cortical areas involved in the cortical-cortical neural oscillatory loop. Cortical-cortical activity is responsible for our conscious experience of reality.

NMDA receptors play a prominent role in layers II and III. NMDA receptors are associated with pyramidal cells.[10] The axons of pyramidal cells extend from the cortex to other areas of the brain. Layers II and III are responsible for cortical-cortical interactivity. NMDA receptors tend to be located on the dendrites of pyramidal cells. In layers II and III, pyramidal cells with NMDA receptors located at different geographical areas of the cortex are involved in an oscillatory relationship.

In combination with neurotrophins, NMDA receptors are critical for cortical-cortical memory. These receptors are involved in both short-term and long-term declarative memory. Short-term declarative memory results from the reinforcement of neural network oscillation as a result of NMDA receptors' participation in the oscillatory process. Short-term memory permits me to remember a phone number for a brief period. In the case of long-term memory, the initial NMDA receptor activity has been further reinforced by activity of the right and left hippocampus. The hippocampus is an important subcortical location of NMDA neuroreceptors. We'll discuss memory in more detail later.

CORTICAL LAYER I

Cortical layer I neurons form a diffusely interactive network whose role in brain function is not well understood. It is suspected that layer I neurons (referred to as the marginal zone in earlier cortical development) reinforce the firing of adjacent columns. This strikes me as a reasonable possibility. Although I've focused upon the individual cortical column for explanatory purposes, it's highly likely that cortical columns located in close proximity to one another fire together. The coordinated activity of adjacent cortical columns would enhance the cortical processing of information.

B. A SUMMARY OVERVIEW

The six-layer cerebral cortex and integration of experience within the brain via neural oscillation patterns are critical to understanding brain function. They are so important that this section serves as an overview summary and further attempt to clarify the central premise of this book. After this overview, we will proceed to part III to discuss psychosocial aspects of brain function.

CORTICAL COLUMNS

The cerebral cortex is arranged in vertical columns somewhat like the hexagonal cells of a honeycomb. Each column contains all six layers of cells. The nerve cells and glial cells of each column are thought to have a common ancestor during early cortical development.

Cortical columns are the functional modules of the cerebral cortex.[11] Within each cortical column, the lower cortical layers (layers IV, V, and VI) interact with subcortical areas of the brain. The upper cortical areas (layers II and III) interact with other cortical areas of the brain (cortical-cortical interaction).

In each column, layers II, III, V, and VI contain pyramidal cells. Pyramidal cells *send their axons to areas outside of its column.* The pyramidal cells of layer VI send axons to the thalamus. Layer V pyramidal cells send axons to subcortical nuclei (other than the thalamus). Layers II and III pyramidal cells send axons to other areas in the cortex. These axons may go to other cortical areas in the originating hemisphere or to the opposite hemisphere.

The neurotransmitter used by these pyramidal cells is glutamate. Glutamate is the primary activating neurotransmitter produced by the cortex. It is rapid acting and utilizes ion channel receptors.

Layers II and IV contain interneurons. These neurons *act within their own column.* Although they may interact with other interneurons, their primary impact is upon the firing of pyramidal cells within the same cortical column. Most *produce inhibitory fields* that influence the firing of pyramidal cells in the cortical column.

The neurotransmitter used by these inhibitory interneurons is GABA. GABA is the primary inhibitory transmitter produced by the cells of the cortex. It is rapid acting and utilizes ion channel receptors. Like glutamate, GABA influences current neural firing without influencing future firing thresholds.

The axons of these inhibitory interneurons are extremely short and begin producing an inhibitory field within the cortical column prior to birth. With cortical development, these inhibitory interneurons are organized into cortical layers II and IV. Layer IV produces the inhibitory field impacting the lower cortical layers (layers V and VI) while layer II produces the inhibitory field impacting layers II and III of the upper cortical layers.

Cortical layers II through VI become functional from the inside out. Layer VI, the innermost layer, is also the only layer that has full functionality at birth. After birth, the remaining cortical layers become functional in a stepwise fashion. Layer V forms reciprocal connections with subcortical nuclei (other than the thalamus). Layer IV develops into a receiving layer for layers V and VI. Layers II and III send pyramidal cell axons to other cortical areas. Layer II develops into a receiving layer for both layers II and III. The outer marginal zone lies over the layers VI through II during their stepwise development. The outer marginal zone becomes layer I of the developed cortex. Layer I may play a role in amplifying and coordinating the activity of adjacent cortical columns.

The frequency of cortical oscillation is increased by reducing the inhibitory fields produced by layers IV and II within the cortical column. The inhibitory fields involving lower cortical layers (layers V and VI) are decreased locally by altering the inhibitory influence of layer IV in response to sensory and other subcortical influences. The inhibitory fields involving the outer cortical layers (layers II and III) are reduced by cortical-cortical activity.

Cortical-cortical neural network is more global in scope. It involves interaction of both sensory and motor areas of the brain. The cortical-cortical neural network is strongly influenced by the anterior portion of the frontal lobe, the prefrontal lobe. The prefrontal lobe directs our attention, focusing on specific sensory input while ignoring others. Concurrent reduction of the inhibitory fields of these two layers (layers IV and II) is associated with gamma wave activity—the frequency associated with conscious mental activity. I suspect that the decrease in the inhibitory field at both cortical areas involved in cortical-cortical neural network oscillation underlies the increase in neural network oscillation frequency to gamma wave activity, which is associated with conscious brain function.

How can we account for a decrease in the inhibitory fields associated with layers IV and II? I can't answer that question as precisely as I'd like. However, there are some interesting possibilities.

Spiny stellate cells are likely candidates. In addition to inhibitory interneurons utilizing the neurotransmitter GABA, layer IV contains activating interneurons named spiny stellate cells. Spiny stellate cells have prominent dendritic fields that spread within layer IV. Their activity is restricted to their cortical column. They are extremely sensitive to incoming stimuli. Thalamic input is received by spiny stellate interneurons in layer IV that have axon fields that reach into layers III, V, and VI. Configured in this manner, spiny stellate cells provide an activating link between thalamic input and cortical pyramidal activation. Spiny stellate cells are "the primary recipients of sensory information received in the neocortex from the thalamus."[12] *The neurotransmitter used by spiny stellate cells is glutamate.*

Chandelier cells in layer IV are also likely to play an important role in modifying the inhibitory field of a cortical column. Chandelier cells have widespread axon fields that interact directly with as many as five hundred pyramidal cell axons (rather than dendrites).[13] Although Chandelier cells are GABA-producing interneurons, they have an activating effect on pyramidal cells. A single chandelier cell is able to activate multiple pyramidal cells at the same time.[14]

Unfortunately, spiny stellate cells and chandelier cells don't account completely for the decrease in the inhibitory fields of layers II and III. Since these cells are found in layer IV, they certainly help to account for a decrease in the level of inhibition in layers IV, V, and VI with subcortical stimulation. They help us to account for the high level of impact that emotional factors have on our perception of reality. I suspect that we'll have to look further to explain the decrease in the inhibitory field in layers II and III required for gamma wave activity between different cortical areas.

MOTOR/SENSORY INTEGRATION

Motor/sensory integration requires interactive connections between motor and sensory cortical areas. At birth, only reciprocal connections between the thalamus and layer VI of the cortex are fully functional. After birth, reciprocal connections between subcortical nuclei and layer V become functional. Reciprocal connections between different cortical areas (cortical-cortical) are last to become fully functional.

Reciprocal cortical-cortical interactive function is initially established between adjacent cortical areas. These involve short cortical-cortical fibers. Over time, cortical-cortical interactive functionality becomes fully established between more distant areas of the cortex. The last of the long cortical fibers to reach full maturity are those that connect the posterior (multisensory) association area with the anterior association area (prefrontal lobe).

Bidirectional interaction between the frontal (motor) lobe and the posterior (sensory) lobes is required for motor/sensory integration. Survival is contingent upon our ability to respond rapidly and appropriately to sensory information. The key to survival is informed behavior.

The prefrontal cortex oversees our voluntary behavior. It assigns priority to sensory input, permitting us to ignore certain sensory input while attending to others. As you're reading this book, the prefrontal lobe maintains your focus on the material and determines when you're ready to move on to the next paragraph. As you're reading, you're able to ignore unrelated sensory information, such as the pressure of your body on the chair, hunger, or nearby sounds. The prefrontal cortex enables you to do this by exerting influence through bidirectional interactivity with the posterior multisensory association area.

The prefrontal cortex and the posterior multisensory association area are at the center of a global oscillating cortical-cortical neural network. The brain's cortical-cortical neural network enables the prefrontal cortex to influence and be influenced by brain activity in the sensory-processing cortical areas of the cortex.

The brain is an integrated, multidirectional neural network. Both the prefrontal lobe and the posterior multisensory association area are continually interactive with other cortical and subcortical areas of the brain. Brain function is the product of massive parallel processing monitoring a wide range of sensory input under the guidance of the prefrontal lobe.

The importance of prefrontal integration of the neural network activity is demonstrated by psychiatric disorders in which this process has been disrupted. Schizophrenia is associated with an overall decrease in frontal lobe activity. Decrease in frontal lobe function is associated with irregularities in sensory perception, referred to as hallucinations. In post-traumatic stress disorder (PTSD), the individual's sensory perception is essentially normal. However, memory of emotionally laden traumatic events can overwhelm prefrontal cortical control, causing the individual to reexperience a stressful event that occurred much earlier.

The loss of prefrontal integration is particularly obvious in individuals with dissociative identity disorder (commonly referred to as multiple personality disorder). These individuals have been unable to develop a coherent sense of self. Instead of a coherent sense of self, a number of identities are formed. We refer to these identities as alters. As you're interacting with the person, you can begin a conversation with one alter and, at some point, be talking to another alter. The transition from one alter to another is often unpredictable. Unless there is a dramatic change from one alter to another, you're likely to be unaware of the shift. Alters are more accurately thought of as personality fragments formed in response to overwhelmingly stressful experiences. Child alters provide a window of insight into the world of alters. Child alters develop during childhood—well before the brain's cortical neural network has reached maturity. A child alter formed in early childhood experiences the world in the same manner it was experienced at that time. A portion of the neural network has become dissociated from the rest. It's as if it were locked in a time capsule. Alters, regardless of the age of formation, can be thought of as dissociated identities that exist independently in a neural network that has been fragmented by trauma.

Brain development requires interaction between the infant and the environment. The most critical participants in the child's environmental interaction are the parents, particularly the mother. The mother provides an interface between the child and the environment. She ensures that the child's exposure to environmental challenges is not overwhelming. As the child's ability to handle environmental challenges increases with age, the mother's level of protection is progressively decreased, permitting the child to become increasingly self-reliant.

The process of brain development depends upon the child receiving sensory input and initiating motor activity. The infant observes the effect of motor activity and modifies motor activity accordingly. This is a feedback process, a concept associated with complex systems that will be discussed further in part IV. The prefrontal cortex orchestrates the process under the influence of the dopamine reward system. With "good enough" parenting (a term introduced by Donald W. Winnecott, a British pediatrician and psychoanalyst[15]), the child develops a stable, positive sense of self and is able to tolerate life's frustrations.

NEURAL NETWORK OSCILLATION

Brain function is the product of neural network oscillation. The brain relies upon patterns of neural network oscillation to model the world we live in. Neural network oscillation enables the brain to provide real-time solutions to complex problems. The actual frequency of neural oscillation is important. The rate of neural network oscillation must be on a par with the speed of the activity being modeled. Radar, for example, must continually sweep the sky at a rate commensurate with the speed of the aircraft it's designed to detect and monitor. If those sweeps were performed hourly, you'd be unable to monitor the path of a slow-flying bird. The brain's neural network oscillation rate must be sufficiently rapid to monitor the types of threats that we're likely to encounter in our day-to-day activity, such as a threatening fist or a falling rock.

Activity that is either too slow or too fast for the parameters of our brain activity is difficult for us to detect. Geological movement is too slow for us to perceive. Prior to modern geology, we tended to assume that our planet was unchanged from time immemorial. A bullet, on the other hand, travels too fast for us to see. There are limits to the range of light and sound that we can detect. We're immersed in phenomena that we can't perceive without sophisticated scientific instruments.

Baseline cortical oscillatory activity is referred to as alpha rhythm. Alpha rhythm refers to the electromagnetic oscillations recorded on an electroencephalogram (EEG) that have the frequency range of 8–12 hertz. Hertz refers to the number of cycles per second. Alpha rhythm is the baseline rate of oscillation in the initial feedback loops between the thalamus and the primary sensory areas. The baseline rate spreads throughout the cortex with cortical development after birth. Alpha rhythm provides the underlying reactivity required for the brain to respond to sensory stimulation. Sensory stimulation focally increases the localized oscillatory rate in the primary sensory areas to 13–30 per second (beta rhythm).

Conscious experience involves the integration of sensory stimuli (e.g., bringing together sight, sound, and touch) into an experience of reality that includes the prefrontal cortex. The prefrontal cortex guides conscious behavior. Integration of cortical processing at this level involves a further increase in the frequency of neural oscillation referred to as gamma waves. Gamma waves have a frequency of 30 to 70 Hz, centered at about 40 Hz. The pattern of neural network gamma wave activity is responsible for our experience of reality and consciousness.

Neural oscillation is necessary for both brain function and brain development. Neurons that fire together wire together. This is referred to as Hebb's hypothesis. Donald Hebb, a Canadian psychologist, proposed in 1949 that the basic mechanism of neural transformation involved increased synaptic efficiency with repeated stimulation across a specific neural synapse.[16]

NEURAL NETWORK TRANSFORMATION

While rapid-acting neurotransmitters are the workhorses of the oscillatory neural network, from earliest infancy and throughout our lives, slow-acting neurotransmitters transform the oscillatory neural network by differentially strengthening synapses in the lower cortical layers in response to our positive and negative experiences. Slow-acting neurotransmitters are produced by subcortical neurons whose axons impact the lower cortical layers.

Slower-acting neurotransmitters work through second-messenger receptors. The neurotransmitter (first messenger) interacts with the external portion of the receptor. The internal portion of the receptor then releases a second messenger. The second messenger goes on to influence some aspect of cell function.

Second-messenger molecules are capable of producing lasting changes in a neuron's function by influencing the cell's protein production. Second-messenger neurotransmission, for example, may increase a neuron's production of neuroreceptors modifying the neuron's future firing threshold. Slower-acting neurotransmission is responsible for our emotional memory.

With further cortical maturation, the outer (cortical-cortical) layers (II and III) become progressively more functional. NMDA neurotransmission plays a prominent role in cortical-cortical memory. NMDA neuroreceptors are often on the same neurons that utilize glutamate neuroreceptors.

NMDA neurotransmission utilizes specialized rapid-acting neuroreceptors that allow calcium ions to enter the nerve cell. Calcium ions act as second-messenger molecules that influence the neuron's firing threshold. NMDA neurotransmission is responsible for our memory of information, which is referred to as declarative memory.

CONSCIOUSNESS: EMERGENT PHENOMENON

Sensory stimulation increases the oscillation rate from baseline alpha (8–12 Hz) to more rapid beta rhythm (13–30 Hz). The EEG activity of the primary visual cortex, for example, shifts immediately from alpha waves to beta waves when we open our eyes. Light waves, sound waves, and somatic sensory stimulation are all translated into beta rhythm neural activity.

Consciousness, as we use the term in everyday language, requires the further transformation of oscillatory activity to gamma rhythm. Gamma waves are taken to be 30–70 Hz, centered about 40 Hz. Consciousness involves concurrent neural network gamma wave oscillatory activity involving the prefrontal cortex and the posterior multisensory processing area along with other sensory and motor areas of the brain.

The pattern of gamma activity is "reality" as we experience it. It reflects a combination of sensory input, prefrontal oversight, and motor cortex activity. With prefrontal focus, areas of beta activity are transformed into the gamma activity and incorporated into consciousness. Prefrontal focus is also able to bring into the network of gamma wave activity memories of previous sensory experiences.

Conscious memory relies upon cortical-cortical NMDA neurotransmission. Emotional memory involving cortical-subcortical neural network interactivity is the product of second-messenger neurotransmission. Conscious memory involves more than our emotional reactions. Conscious memory is involved in information recall. Conscious memory relies upon cortical-cortical interaction involving cortical layers II and III. In the adult neocortex, NMDA receptors are located primarily in layers II and III.

In addition to NMDA receptors in layers II and III of the neocortex, NMDA receptors are also located in the subcortical hippocampus. The hippocampus promotes memory by reinforcement of gamma wave activity. Evidence suggests that the hippocampus has an internal oscillatory rate of forty cycles per second.[17] An internal oscillatory rate of forty cycles per second would enable the hippocampus to work synergistically with and reinforce gamma wave activity (which is also in the range of forty cycles per second). Without participation of the hippocampus, the brain is unable to reinforce patterns of neural interaction and form long-term memories. A person with damage to the hippocampus appears to act normally but can't remember events that took place a short time earlier.

The role of slow wave (theta, around 4–7 Hz and delta, less than 4 Hz) activity is not clear. Slow wave activity is often associated with pathological brain function. It has been suggested that theta activity is related to shifts in brain activity from beta to gamma wave activity.

PART III ─────────────────
PSYCHOSOCIAL ASPECTS
OF BRAIN FUNCTION

A. EXPERIENCE SHAPES NEURAL NETWORK

At birth, the neural network is primarily a product of genetic evolution. After birth, the day-to-day function of our brains is progressively influenced by experience. Our relationship with our parents has much more influence over our moral and language development than genetic inheritance.

The brain learns by strengthening neural connections. Language fluency, for example, involves a strengthening of neural network connections based upon repeated language experience. It's a form of overlearning through repeated verbal interaction with others. English language patterns have been overlearned to the point that I can anticipate an English speaker's next word or phrase. As I'm speaking, I'm able to sort out the precise phrase needed to complete a sentence. Although I've studied other languages in school, I don't have anything approaching the fluency I have in English.

The brain is biased toward reliance upon overlearned (strengthened) neural connections. Our sense of self and reality are forms of bias in our perception associated with overlearned (strengthened) neural connections. We perceive ourselves and the world around us in a fashion that has been shaped by our life experiences.

Psychological and social aspects of brain function are dependent upon our experiences. The dots of our experiences can be arranged in many possible stable variations of neural network interaction. What works in one society often won't work in another. The psychological and social aspects of brain function are relative, not absolute, phenomena.

As I discuss the psychological and social aspects of brain function, I'll refer to concepts introduced by a variety of authors. Most are contributors to the literature of psychiatry and psychology. Some represent more philosophical or religious perspectives. By referring to the work of such authors (e.g., Freud, Jung, Piaget, Buber, and others), I'm hoping to demonstrate that the explanation of brain function I've provided is consistent with the observations of others who've wrestled with these issues.

NARRATIVES

Language involves sounds that have a socially shared meaning. If I say the word "table," another English-speaking person will understand what I'm referring to. We bring together (chunk) a variety of information into words. The word "table," for example, brings to mind (chunks) a variety of features that we associate with tables, such as a flat surface raised above floor level, often on four legs, and convenient for placing items.

We use language to construct narratives. The word "table" isn't used in isolation. It is part of a narrative formed by bringing words together. Language permits us to construct a shared understanding of the world around us.

Narratives connect the dots of our experiences into a conceptual framework. Social groups define themselves through narratives. Every society connects the dots of their experiences through their narratives.

We have a strong bias toward a linear organization of information. By linear, I mean that we organize information in a time-based fashion in which events are associated with preceding events and events that follow. Linear organization of information reflects our personal experiences. As we go through life, every event is associated with those that precede it and those that follow. Our bias toward linear organization of information is reflected in our narratives which routinely present phenomena as a time-ordered sequence of events.

However, complex phenomena don't lend themselves to linear analysis. Complex phenomena are the product of multiple factors acting concurrently, often in an unpredictable manner. Different groups observing the same complex phenomenon will construct different linear narratives to explain what took place. None of those linear narratives is capable of capturing the full complexity of the phenomenon they describe.

Without actually understanding complex phenomena, we often use metaphors to describe them. Metaphors are a type of narrative that likens an unfamiliar complex phenomenon to one that is more familiar. The kingdom of God, for example, may be likened to a mustard seed that grows into a magnificent shrub.

Metaphors provide a false sense of understanding. I'm able to note similarities between a familiar complex phenomenon and one that is less familiar without actually understanding either. From the earliest times, we've used metaphors as a substitute for understanding complex phenomena.

Our narratives also tend to be self-serving. They provide us with a sense of security and bolster our self-esteem. A hunter-gathering tribe, for example, will have its own narratives that stress the importance of the tribe and tribal traditions. Typically, the tribe sees itself positively and outsiders less positively, often negatively.

The same is true of religious groups. Each tends to regard its own beliefs as truth and the beliefs of others as error—manifestations of evil in the world. The importance of religious narratives to our sense of well-being is highlighted by the frequency of religious warfare in which each side is convinced that its cause is morally correct. In times of stress, instead of reexamining our narratives, we cling to them even more tenaciously.

Our brains create narratives by connecting the dots of our experiences into meaningful patterns. They do so by strengthening some connections while others are weakened. The brain connects the dots of our experiences in a manner that biases our reality to enhance our sense of security.

Narratives evolve in a stepwise, incremental fashion. Once formed, they tend to remain stable and to resist change. We cling to our narratives because change provokes anxiety. Change tends to occur only after a substantial body of conflicting evidence can no longer be ignored. We refer to such changes as "paradigm shifts."

MEMORY

The creation of narratives is dependent upon declarative memory. Declarative memory refers to our ability to recall, organize, and express past experience. In most cases, it refers to long-term memory—our ability to recall events that happened hours or years ago. Declarative memory is conscious memory.

Declarative memory relies upon NMDA neurotransmission. As we discussed previously, NMDA (shorthand for N-methyl D-aspartate) receptors are a type of glutamate rapid-acting neuroreceptor. Glutamate rapid-acting neuroreceptors and NMDA receptors tend to be located on the same neurons in layers II and III. Like other rapid-acting neuroreceptors, NMDA receptors utilize an ion channel that permits positively charged sodium ions to enter the neuron.

Unlike glutamate rapid-acting neuroreceptors, NMDA neuroreceptors also permit calcium ion to enter the neuron. Inside the nerve cell, calcium ions function as a second messenger that creates memories by influencing the neuron's future firing threshold. In addition, NMDA neuroreceptors are voltage-dependent. Their ion channel contains a magnesium ion that blocks the free flow of ions. The magnesium ion pops out only when the neuron membrane is depolarized. NMDA neuroreceptors don't permit ion flow until the neuron is actually depolarized.[18]

In the cortex, NMDA neurotransmission is located primarily in the cortical-cortical layers (II and III). NMDA neurotransmission in the cortical-cortical layers is critical for the formation of declarative (conscious) memory. Emotional memory, on the other hand, is dependent upon slower-acting second-messenger neurotransmission produced by subcortical nuclei. Slower-acting second-messenger neurotransmission modifies the frequency of interactive oscillation between subcortical nuclei and the lower cortical layers.

In addition to cortical NMDA receptors, *long-term* declarative memory requires the hippocampus. The hippocampus is an important subcortical location of NMDA receptors that interacts widely with the cerebral cortex. It is critical for long-term declarative memory formation.[19] Should the hippocampus be destroyed bilaterally, new long-term memories cannot be formed. Under these circumstances, declarative memories can be retained for only a short period.

Emotional memory, on the other hand, is independent of the hippocampus. If your right and left hippocampus were destroyed, I could enter a room, talk to you, leave, and come back a short time later without your having any conscious memory of my visit. However, if I kicked you in the shin before I left, you'd experience an emotional response when I returned, without being able to explain why. The inability to integrate emotional memory and declarative memory is thought to be important in post-traumatic stress disorder (PTSD).

Declarative memory involves the strengthening of cortical-cortical neural network oscillatory patterns. Repetition fosters memory through NMDA receptors together with neurotrophic factors.[20] Practice does make perfect. Neural transmission is strengthened by increasing the density of synaptic connections, which is dependent upon the availability of neurotransmitters and the density of neural receptors.

Neurotrophic factors strengthen synaptic connections between neurons that fire together. This mechanism contributes strongly to the formation of oscillatory neural networks and their later myelination. Neurotrophic factors are important for strengthening synaptic connections prior to birth and throughout life.

NMDA neuroreceptors in cortical layers II and III of the mature cortex reinforce the firing of neurons involved in gamma wave activity. Reinforcement of gamma wave activity permits short-term declarative memory. Short-term declarative memory enables me to remember information for the moment without recollecting it later.

Long-term declarative memory requires cortical-cortical connectivity bolstered by hippocampal reinforcement. The hippocampus may have an internal oscillation rate of approximately 40 Hertz that reinforces gamma wave activity (also about 40 Hertz).[21] The hippocampus has an exceptionally high metabolic rate that renders it even more sensitive to oxygen deprivation than other areas of the brain.

MYELINATION AND CORTICAL MATURITY

As the brain develops, cortical processing is enhanced by axon myelination of major axonal pathways. Myelinated axons permit geographically separate portions of the brain to interact more rapidly with one another. I strongly suspect that myelination is directed toward repetitively firing neural axons—those involved in neural oscillatory activity. Myelination is the last step in the maturation of the brain. With myelination, the individual sensory processing areas and the posterior multisensory area are able to function in an integrated fashion. Myelination also enables the prefrontal lobes to function in an integrated manner with the posterior multisensory association areas in the parietal lobes. The longest myelinated fibers connect the prefrontal lobe with the posterior multisensory association area. These are the last to fully myelinate. Myelination of the prefrontal lobe is not complete until late adolescence or early adulthood.

DEVELOPMENTAL COURSE

The developmental course of human brain development echoes the observations of Jean Piaget (1896–1980), a Swiss developmental psychologist. He postulated four major periods of cognitive development: the sensorimotor period (birth to about age two), the preoperational period (age two to about age seven), the concrete-operational period (age seven to about age twelve), and the formal operational period (age twelve and on). The sensorimotor period is influenced primarily by the connectivity of layers V and VI of the cortex. The preoperational period may be thought of as reflecting a transitional phase between sensorimotor and cortically directed activity. The operational period reflects the maturation of cortical-cortical activity. Piaget believed that the child progressively develops cognitive structure in a stepwise fashion that involves the "assimilation" of new information into the existing cognitive structure and periods of "accommodation" in which cognitive structures are modified to accommodate new information.[22] The stepwise fashion in which a child's processing of information develops is somewhat similar to the stepwise progression (paradigm shifts) involved in the evolution of narratives.

IMPORTANCE OF ENVIRONMENTAL INTERACTION

Ongoing interaction with our environment is critical for brain development. If a newborn cat is blindfolded for the first six months of life, it will never develop the ability to process visual stimuli. The cat's visual interaction with its environment is needed for those aspects of brain function to develop. There is a critical period during which this must occur.

A newborn with cataracts is unable to react to light. If cataract removal is delayed past a critical period, the child will never learn to see. The primary visual processing area of the brain won't develop in a normal manner without visual interaction with the environment.

Normal adult brain function continues to require ongoing environment interaction. Sensory deprivation may produce anxiety, difficulty concentrating, even hallucinations and delusions. Individuals who lose sight or hearing may become paranoid due to their inability to accurately perceive what's going on around them.

B. REALITY

Reality is the product of brain function. Our brains construct the world around us as they process our sensory perceptions. As we go through life, our reality is fashioned through the feedback interaction involving our sensory perception (including our emotional reactions), prefrontal processing, motor activity, and perceived impact on the environment.[23]

Reality differs from one species to another. It reflects the specific sensory input and associated patterns of brain function characteristic of each species. We have taste, smell, hearing, vision, and the somatic sensory input, such as pain and touch. Other species have different sensory parameters. Bats and dolphins rely on a form of sonar. Sharks are sensitive to their electrical environment. Migratory birds are sensitive to the magnetic environment. Their reality is quite different from the reality that you and I experience.

There is a pragmatic relationship between the reality that we experience and the world around us. The brain's approach to information processing would have little survival value if it did not represent the world around us in such a way as to promote survival. Nonetheless, reality and the world around us are not the same.

The brain emulates the world around us. It creates an internal version of the external world. We're able to rapidly manipulate reality produced by the brain and use this skill for problem solving and guiding our behavior from moment to moment.

The key to the speed of the brain's processing of complex information is the speed of neural conduction. The living brain's electrical activity is continuous. It forms a dynamic medium that responds almost instantly to environmental stimuli.

PATTERN RECOGNITION

The brain relies upon overlearned patterns of activity. The printed word "stop" is an overlearned pattern for English-speaking readers. Even if portions of the word are covered, you're still likely to recognize it.

The brain is sensitive to overlearned patterns and recognizes them quickly, even with only limited exposure. Overlearning is the key to language processing. Due to overlearning, I'm able to rapidly recognize your meaning (if you're speaking English) and can begin to formulate a response as I listen. Language fluency is a form of overlearning. Without fluency, I'm limited to processing your words one at a time. You've moved on to the next sentence before I've sorted out the first phrase.

Pattern recognition enables us to make surprisingly accurate assessments with only incomplete information. We're able to recognize a familiar person at a distance based on his or her gait and posture. Surrounded by people talking, we're able to recognize a specific person's voice based on tone and inflection. We're even able to focus on that voice while ignoring others.

The brain has a bias that causes it to move automatically toward overlearned associations. For example, the color yellow and the word *banana* are closely linked in our experience. When either *yellow* or *banana* comes to mind, it evokes the other.

Our ability to accurately process complex patterns of information with only partial data has enabled us to survive as a species. Our brains effortlessly handle the ambiguity encountered in day-to-day existence. In sharp contrast, our brains didn't evolve to do calculus or to understand the theory of relativity. Those activities aren't effortless at all.

Without specialized training, you and I can throw a Frisbee back and forth. Our brains are able to process the flight path of the Frisbee and rapidly initiate the behavior needed to catch it. Catching and throwing a Frisbee is well within the parameters of the brain's processing ability, which has evolved to deal with the day-to-day events that we encounter. These are the events that we see, hear, and feel. We've learned through scientific discovery that there are a wide variety of phenomena beyond the brain's ability to process that are outside of our day-to-day awareness.

We resist changes to overlearned patterns of brain function. Our sense of stability and security is strongly related to them. The military, for example, is often accused of preparing for the last war—rather than future wars, which are likely to be fought differently. Attempts to introduce a paradigm shift tend to fall on deaf ears. Moving beyond the accepted and familiar creates uncertainty and insecurity. We tend to protect ourselves from insecurity and anxiety by not seriously considering opposing views. "I may be wrong, but I'm never in doubt."

PROBABILITY-BASED PROCESSING

Brain processing is probability based. The adult mind doesn't process information de novo. Every perception is influenced by prior experience.

Probability-based processing is the outcome of pattern recognition. Every new perception is interpreted in terms of the patterns of brain organization that have already been established. Our current perceptions are influenced by preexisting steady-state brain patterns. "Steady state" is another term associated with complex systems and will be discussed further in part IV.

This approach has been referred to as Bayesian. The Bayesianism is based upon the work of Thomas Bayes (1701-61), an English clergyman whose work was published in 1764. Bayes described a method of combining expectations based upon earlier experience with new perceptions to produce an accurate guess concerning the probable significance of the new information.[24]

Reality is a pattern of neural network activity. Our previous experience lays the foundation upon which our current experience takes place. The brain's processing of information is probability based. Reality is progressively modified by new experience.

SELF

Our sense of self refers to how we experience ourselves. Our sense of self, our goals, our values, and our judgments of others are shaped by positive life experiences. Positive experiences are those that we'd like to repeat (while avoiding negative experiences). Our sense of self gives us a sense of stability and certainty as we confront life's uncertainties. We protect our sense of self by resisting change. Changes normally occur incrementally.

The prefrontal lobe is a key player in the development of a sense of self and in guiding our behavior.[25] Injury to the prefrontal cortex is often associated with disruptive personality changes. Keep in mind that the dopamine reward system impacts only the frontal lobe, particularly the prefrontal cortex. Our sense of self, shaped by the brain's dopamine reward system, is narcissistically biased.

Due to the influence of the brain's dopamine reward system on the frontal lobe, there is a systematic skewing of our perception of the world around us. We tend to see the world in terms of good and evil. Due to our narcissistic bias, we are more likely to recognize good in ourselves and evil in others. There is often animosity and distrust between social groups. In virtually every war, each side is convinced that it represents the forces of good while the opponent represents those of evil.

Excessively negative experience threatens the integrity of our sense of self. I'm not suggesting that life should be limited to a series of positive experiences. Frustration and negative experiences are an inevitable part of life. We need to learn to adapt to frustration. Tolerance for frustration and negative experience develops gradually. At birth, a child doesn't have the mental capability to tolerate frustration. Over time, the child slowly develops a sense of self and the ability to tolerate frustration to gain social approval.

Good parenting involves progressive modification of the degree of support provided, depending upon the child's phase of development. Ideally, parents provide sufficient support to enable the child to maintain a stable sense of self in the face of life's frustrations. They protect the child from frustration that he or she is unable to tolerate. If children experience more frustration and negative experiences than they can tolerate, their sense of self may be disrupted. In severe cases, children may develop multiple personality disorder, and adults may develop PTSD. Good parenting involves protecting the child from overwhelming frustration rather than from all frustration.

As the child matures, the prefrontal lobe, allied with the brain's dopamine reward system, becomes increasingly able to influence our focus of attention and behavior. The developing self seeks to maximize positive experiences by focusing our attention and behavior to social expectations. This process is referred to as character formation. Every society attempts to influence character formation in such a way that the individual will reliably contribute to the community enterprise.

The self is an attractor system—a relatively stable configuration of neural network activity. Attractor systems are patterns of activity that evolve over time in response to multiple variables. They are a feature of complex systems. We'll review attractor systems further in part IV.

Attractor systems vary in their stability and may react unpredictably under certain circumstances. With adequate positive developmental experience and appropriately incremental exposure to the unpleasant aspect of life, our children will develop a stable sense of self that resists the adverse impact of stress. On the other hand, if our children are overwhelmed by traumatic experiences, their sense of self will tend to be a weaker attractor that's more easily overwhelmed by future stress.

SHARED REALITY

Reality is a private experience. The young child's worldview is quite unique and (from an adult perspective) unrealistic. As the years go by, children progressively adopt the worldview of their parents and other adults in their society.

Language and communication enable us to move toward a

shared view of reality. Through verbal interaction, we are able to move beyond the private world of childhood to the shared worldview of adults. A shared view provided through language enables human beings to function in social units.

Our sense of self is highly dependent upon language and our interactions with others. We define and experience ourselves in terms of our relationships with others. Our sense of self is strongly influenced by how others react to us. Language permits us to develop a sense of self that incorporates verbal feedback from those around us.

Culture refers to the shared values, roles, and language by a social group. It is the product of language and communication. Language enables us to share our experiences with one another and to pass them on to following generations. Culture enables us to function effectively in social groups over multiple generations.

Culture includes shared concepts about the nature of reality. As societies evolve, shared concepts inevitably change—paradigm shifts occur. In our times, paradigm shifts are taking place with increasing frequency due to advances in science and the transition to a global socioeconomic network.

Discontinuities develop between and within societies. As we'll see in part IV, discontinuities are periods of unpredictability that are characteristic of complex systems. Discontinuities between societies can result in war. Within a society, social elements that are isolated from mainstream social development tend to become seedbeds of discontent and social conflict.

C. HUMAN ADAPTATION

Human adaptation relies upon a feedback process. Our senses provide information about our surroundings. Our prefrontal lobe acts upon that information in a goal-directed manner. A favorable response leads to repetition. An unfavorable response discourages repetition. This feedback process shapes the connections of the brain's neural network.

The ability to discriminate positive from negative outcomes is critical. Early in life, we respond to short-term outcomes based upon pleasure and pain. Sigmund Freud (1856–1939), the founding father of psychoanalysis,[26] referred to this as the pleasure principle.[27]

Prefrontal lobe development enables us to take long-term consequences into account. Freud referred to this as the reality principle.[28] It requires a vision of where one wishes to go and the ability to reach back in memory for information concerning possible outcomes of different actions. This ability is markedly enhanced by the use of symbol systems, such as language (both spoken and written). With symbol systems and the ability to communicate, we are able to look much further back into the past and to anticipate outcomes further into the future.

Human adaptation is proactive. We try to anticipate problems and to avoid them. If we can't avoid a problem, we try to find a solution before the problem becomes completely unmanageable. We carry in our head a vision of the type of future we seek to achieve, and we work toward it.

Problems in human adaptation occur at both the individual and social level. An individual may be psychotic or too depressed to function. Psychiatrists have diagnostic terms for these problems. However, adaptive problems may reflect broader societal difficulties. In modern western society, there is often little sense of community, with many individuals feeling socially disconnected and insecure.

APPROACH/AVOIDANCE

The nucleus accumbens is a key component of the brain's dopamine reward system and plays a prominent role in positive motivation. It motivates us to seek out ways of meeting our needs. The nucleus accumbens, as a component of the basal ganglia, shapes voluntary movement toward rewarding behavior.

At the cortical level, the dopamine system interacts with only the frontal lobe, which is responsible for voluntary movement under the direction of the prefrontal lobe. The nucleus accumbens has such a powerful influence on behavior that a rat wired to electrically self-stimulate the nucleus accumbens will starve to death. It won't stop self-stimulating the nucleus accumbens long enough to eat. Cocaine works through this system.

The amygdala plays a key role in our avoidance system. The amygdala is actually a cluster of nuclei that utilize several neurotransmitters, both rapid-acting (including NMDA) and slower-acting neurotransmission. It has reciprocal pathways throughout the cerebral cortex. The amygdala is also closely connected to the hippocampus.

The amygdala and the nucleus accumbens have a pervasive influence upon cortical development. They are reciprocally connected to the layer V pyramidal cells. The neurotransmitters of these subcortical nuclei utilize second-messenger neuroreceptors that have a long-term modulating effect on cortical function.

Our early experience is sorted out by positive and negative emotional valence. Young children tend to see everything in black-and-white terms. They are not capable of integrating these differing experiences into a coherent experience of themselves or others. We maintain aspects of this early dichotomy throughout life.

Our positive experiences serve as a foundation for our sense of self. We tend to see ourselves and our own social group in a positive light (while tending to see others in a more negative light). We commonly divide the world into good and evil.

ROLE OF SECURITY

The need for a sense of security underlies human behavior. Security involves more than physical safety. Early in life, our sense of security is threatened by parental disapproval and enhanced by parental approval. During the teen years, our sense of security is threatened by peer disapproval and enhanced by peer approval. As adults, our sense of security is related to our sense of competence and control. Even a competency-based sense of security is enhanced by the approval and respect of those whose opinions we value. The need to be valued by others is important throughout the life cycle.

Beginning in infancy, slower-acting neurotransmitters (those involved in emotional memory) systematically strengthen neural connections that enhance security. Neural connections that promote safe behavior and prevent unsafe behavior are shaped by experience. They serve somewhat like a ship's rudder that turns us toward behaviors that enhance physical and social security and away from behaviors that result in a sense of insecurity.

Dopamine, norepinephrine, serotonin, and acetylcholine (referred to as biogenic amines) are slow-acting cortical neurotransmitters. They are relatively simple molecules that neurons can produce and transport easily to vacuoles in their axon tips. Neurons produce neurotransmitters from even simpler, readily available components.

The biogenic amines produced by subcortical neurons interact with second-messenger neuroreceptors in the lower layers of the cortex. In this fashion, neural connections in the lower cortical layers are modified in response to emotional reactions. Cortical-cortical neural network activity, involving the upper cortical layers, matures later and builds upon this foundation. Emotional and declarative memory work synergistically to create our sense of good/evil, our sense of self, our perception of the world around us, and our goals—all contributors to our overall sense of security.

ROLE OF DELUSION

Human reality has a strong delusional component. Our belief systems enable us to maintain a manageable level of anxiety despite the uncertainties of human existence. Our sense of self, for example, is the product of our positive experiences. We strive to maintain a sense of specialness. We tend to overlook the less flattering aspects of our own behavior although we can identify flaws in others with relative ease.

Delusions have played an important role in human survival. They have enabled us to carry on despite adversity. Protected by the belief that God is on their side, soldiers in combat are better able to carry on despite the imminent threat of death. Soldiers of every army and every religious persuasion have faced danger more confidently with such beliefs.

Jules Masserman (1905–94), an American psychiatrist, proposed in the mid-twentieth century that humans "defend against psychological disorganization and traumatization by three basic Ur-defenses: belief in their physical invulnerability, the fantasy that other humans are potential friends and helpers, and faith in a celestial order."[29] These Ur-defenses enable us to maintain a personal sense of security despite the genuine threats of injury or death that we face throughout life.

The Ur-defenses reflect a skewing of experience through an emotional prism. Our perception of reality is shaped by our reward/avoidance system. We identify with the positive and distance ourselves from the negative.

Ur-defenses provide a delusion of security. They enable us to function without being overwhelmed by anxiety. We cling to them tenaciously. However, even the Ur-defenses can be overcome by life's circumstances. Post-traumatic stress disorder, for example, is associated with a breakdown of this delusional system.

We find it tragic when our delusion of security is overcome by reality. How can bad things happen to good people? We tend to explain the occurrence of bad things by blaming the victims—it must have been brought on by their misbehavior. By blaming the victim, we're able to maintain our own sense of security—we don't have anything to worry about because we have not engaged in their misbehavior.

Comedy, on the other hand, pokes fun at the delusions of others without threatening our personal sense of security. It's much easier to find humor in the beliefs and behavior of those we regard as different from us. When it strikes closer to home, we're more likely to experience a sense of tragedy.

The Ur-defenses play a prominent role in our religious traditions. They permit believers to face the ultimate threat to personal integrity, death, with some modicum of equanimity. Many early Christians welcomed death by martyrdom in a Roman arena. The same can be said for Islamic suicide bombers today. Religious traditions have contributed strongly to our sense of personal and social security. However, differences between religious traditions can pose serious stumbling blocks to social harmony as the human community becomes increasingly global.

MAGICAL THINKING

Delusional beliefs are often supported by magical thinking. Magical thinking refers to the incorporation of clearly unrealistic factors to support our narratives. Magical thinking is common in fairy tales. Cinderella's life is transformed from one of thankless drudgery to that of a princess through the magical intervention of her fairy godmother. Jack's life is transformed from that of a boy with poor judgment to a hero of legend through magic beans that give him access to another reality.

Greek plays had the option of salvaging the situation by having an actor portraying a god descend to the stage on a machine (deus ex machina) and bring the play to a satisfying conclusion.

God plays that role in our religious systems. God provides an explanation for anything we're unable to explain. How was Earth created? God made it. What will happen tomorrow? God will provide. God provides a fudge factor that enables us to connect the dots of reality into a pattern that promotes security.

Magical thinking is common in our day-to-day lives. We anticipate magical breakthroughs in technology that will eliminate life's problems. We assume that human genius will overcome all obstacles. We believe that democratic elections will lead inevitably to positive social transformation. We act as if every country can be transformed into a successful democracy by introducing elections, ignoring the complexity of political reality.

BIOLOGICAL BIAS FOSTERS SOCIAL COHESION

Socialization refers to the shaping an individual's sense of self, use of language, values, and understanding of the world. Socialization is the product of a child's relationship with their immediate family and the larger community. It involves a series of repetitive activities that shape the neural network connections. Socialized individuals have taken on the beliefs and skills needed to function in their society.

Societies promote social cohesion through the socialization process. Social cohesion involves a supportive relationship between members of a social group. Human social cohesion is based initially on family relationships. Historically family relationships evolved into tribal relations. Over time, much larger social units have evolved. In general, these larger social units have a significant evolutionary advantage. The sheer number of people in a modern nation state tends to engulf and incorporate smaller tribal societies.

Shared language is particularly helpful for the cohesive function of social groups. Shared language enables groups to work cooperatively toward shared goals. Without shared language, construction of the Tower of Babel came to a standstill.

Religious beliefs have contributed strongly to social cohesion. The Christian tradition directs us to "love thy neighbor as thyself." Every major religion encourages us to value our fellow human beings (especially those who share our religious beliefs).

In addition, however, our brains are biologically biased toward social cohesion. By nature, we're attentive to the reactions of other human beings—particularly our caregivers during infancy and childhood. Infants learn to speak by listening to and imitating their parents' behavior.

Humans of every society foster social relationships by systematically avoiding disruptive behavior and acting in a socially acceptable manner. We refer to such behavior as "polite." We're very aware of others' reactions to our behavior and systematically attempt to avoid antagonizing individuals whose relationship we value.

The neurotransmitter serotonin plays a role in social cohesion. Serotonin is another second-messenger neurotransmitter produced by subcortical nuclei—in this case, the raphe nuclei. The raphe nuclei are a cluster of neurons that lie below the thalamus in the brain stem. They are an even earlier product of our evolutionary history than the nucleus accumbens and amygdala that lie in the limbic area surrounding the thalamus. The serotonin system projects diffusely to the more recently evolved areas of the brain where it helps regulate wake-sleep cycles, affective behavior, food intake, and sexual behavior.

Serotonin-specific reuptake inhibitors (SSRIs) are medications used to modify serotonin activity. SSRIs inhibit the reuptake of serotonin after it has been released into the synaptic cleft. The overall activity of the serotonin system is modified in this manner. Prozac was the first of the SSRIs to come on the market.

Dr. Peter Kramer, in his book *Listening to Prozac*, indicates that Prozac is very helpful for anxiety and depressive symptoms associated with "rejection sensitivity."[30] In my own practice, I found his analysis very helpful. Many of my anxious and depressed patients appeared to be reacting to rejection sensitivity and benefitted from treatment with an SSRI.

The need to be accepted and valued by others plays a central role the process of socialization. Children learn to conform to parental expectations in order to maintain a secure sense of being accepted and valued by their parents. They automatically take on their parents' values. The same process continues during school years and into adulthood. The army and every organization that requires teamwork rely upon the basic tendency of human beings to strive to be a valued member of their social system.

Social sensitivity has both positive and negative consequences. On the one hand, it permits people to function as social units. Social units, beginning with the family, are critical to human survival. However, as the human population grows and becomes more diversified, we can't belong to every group.

Rejection sensitivity strengthens group cohesiveness while magnifying differences between members of different groups. We see this phenomenon in high school cliques, in street gangs, among racial groups, and between nations. While social cohesiveness maintains social structure, it also threatens to undermine human existence as different social groups are increasingly required to interact with one another as a result of technological advances in travel and communication.

As the world has become increasingly interactive, we're forced to deal with conflicting belief systems. It's becoming much more difficult to rest comfortably in the security of an isolated cultural tradition. We must find some way to accommodate one another.

Historically, large expanding social systems have simply overwhelmed and assimilated smaller social groups. The Roman Empire assimilated tribal societies throughout Europe. The Roman military viewed themselves as the bearers of civilization to the barbarians. Over time, Roman Peace (Pax Romana) spread throughout most of Europe. Tribal gods were exchanged for Roman gods.

Today, there are a number of large social systems competing for the earth's resources. Through advances in communication and transportation, these large social systems are increasingly interactive with one another. Our economic community has become virtually global in terms of ideas and commodities. Interaction between the large social systems of today is inevitable. However, these large social blocs have been unable to simply overwhelm and assimilate each other.

It is particularly important at this phase of cultural evolution to gain a better understanding of human brain development and function. Failure to recognize the delusional skewing that is inherent in brain and social function threatens human survival. Cultural accommodation is virtually impossible when every party involved is convinced that its own cultural or religious tradition represents truth while all others are evil.

FUZZY LOGIC, CERTAINTY, AND SECURITY

Fuzzy logic permits us to maintain a sense of certainty despite the day-to-day uncertainties of life. I'm referring to the certainty that every social group has in its relationship with the gods, in the importance of their own social group, in the benevolence of the universe, and confidence that human effort will find a way forward. Without certainty in these areas, existential anxiety (the anxiety associated with full awareness of the day-to-day threats to our existence) could easily be overwhelming.

My own exposure to fuzzy logic is through the book *Fuzzy Logic* by Daniel McNeill and Paul Freiberger.[31] McNeill and Freiberger provide a readable account of the thinking of Lotfi Zadeh, who introduced the concept of fuzzy logic. Concepts associated with fuzzy logic may help to clarify the position I've outlined in this book.

Fuzzy logic maintains that human thought is imprecise. Human thought relies upon language. Language is the product of social consensus. It is a tool of individual and social survival. It is not a tool of scientific precision.

I agree. Brain function is inherently imprecise. It is the product of massive neural network parallel processing. McNeill and Freiberger point out that our experience involves networks of neurons that are firing together. Our memories involve groups of neural network cellular assemblies[32] that have fired together in the past and are likely to fire concurrently when portions of the network are stimulated. There is a high degree of variability between patterns of neural network activity from one individual to the next.

Despite the imprecision of brain function, we strive for the security of certainty. We seek a sense of certainty regarding issues for which we have no firsthand knowledge, such as the creation of human beings and life after death. We require certainty for a sense of security.

Our need for certainty and security reflects the motivational underpinnings of human behavior. Our dopamine reward system leads us to seek certainty despite life's uncertainties. Even the most primitive social groups have evolved beliefs that enable them to face life's challenges with a sense of certainty. In every war, people on all sides are convinced of the righteousness of their causes. Both sacred scripture and modern science reflect man's striving for certainty.

Our need for certainty is reflected by the persistence of paradigms.[33] Instead of regarding science as the rational accumulation of truth, Thomas Kuhn described it as a series of paradigm shifts.[34] Each paradigm unites the body of existing information into a pattern of explanation that we resist changing. The concept of a spherical Earth initially was met with resistance. The notion that Earth is not the center of the universe was also met with resistance. The notion that human life has evolved on Earth rather than being created de novo continues to meet resistance.

Even our scientific paradigms are pragmatic approximations of the "truth." They are tools that we have developed in our quest for a security in a universe that is indifferent to our survival. Hopefully, they will provide increasingly more accurate models for understanding ourselves and the universe we live in.

Brain function is inherently pragmatic. McNeill and Freiberger indicate that "the brain has between 10 and 100 billion neurons, and each can have 1,000 to 5,000 synapses."[35] The pattern of neural network activity depends on the relative strength or weakness of the connections involved. There are innumerable possibilities. Of those possibilities, the key issue is whether or not that pattern of neural network activity promotes the survival of the individual and of the social group.

The neural network pattern of cortical-cortical connections is founded upon the previously established cortical-subcortical network. The cortical-subcortical network is shaped by our approach-avoidance emotional base of experience and our need for a sense of security. There are innumerable possible cortical-cortical connections that can be established. We cling to those that provide us a sense of security.

The inherent imprecision of cortical-cortical neural network interaction (as reflected in fuzzy logic) permits the evolution of a wide range of emotionally invested conceptual frameworks. Many different religious systems and nation states are made possible by the variety of neural network configurations the human brain is able to create. Once established, we cling to these systems that shelter us from the life's uncertainties.

CREATIVITY AND THE PREFRONTAL LOBE

Charles Darwin's work illustrates the process of creativity. In the mid-1830s, Darwin, a naturalist on the HMS *Beagle*, was struck by the differences in species from one island to another in the Pacific Ocean Galapagos Islands. In 1859, he published *On the Origin of Species* in which he proposed that species are modified over time by a process of natural selection,[36] often referred to as "survival of the fittest."[37] At the time, species' modification over time conflicted with the widely held belief that God created the species as they exist today. The implication that human beings are the product of species modification was particularly unwelcome.

Charles Darwin questioned the more traditional explanation of species origin after observing the subtle differences in specific species from one island to another within the Galapagos. He also noticed that the creatures on each island appeared well adapted to their specific environments. And he wondered how these differences came to be. He set aside the traditional explanation and considered other possible ways to connect the dots in response to a series of new observations. By delaying closure on the issue of species differentiation, he was able to provide a novel explanation that better accounts for his observations.

Creativity involves finding a novel way of connecting the dots. Fuzzy logic lends itself to a wide range of ways in which the dots of our observations may be connected. Virtually every society, from the hunter-gatherers to modern societies, has connected the dots in such a manner as to gain a sense of understanding of its surroundings. Every new scientific observation challenges how we connect the dots today.

The brain connects the dots by strengthening some neural network associations while others weaken. Language development illustrates the strengthening of neural connections. Our experience becomes closely associated with the words and phrases of our native language. In my native language, the order and phrasing of words are overlearned to such an extent that I can anticipate how a sentence will develop after hearing the initial phrasing. I can begin to formulate a response before you've even completed your statement.

Creativity involves both novelty and utility. A psychotic individual connects the dots in a manner that is much different from you or I. Novelty alone is not necessarily useful in human society. Novel and useful ways to connect the dots are those that survive over time because they enhance survival of a society in one way or another.

Utility is relative. The gods of a hunter-gathering society contribute to the group's social cohesion as it interfaces with the surrounding world. In the absence of contact with more powerful social groups, they have little reason to question the validity of their gods. Anyone who questioned those beliefs would be perceived as a social threat and dealt with accordingly.

In a similar manner, Darwin's concepts in *On the Origin of Species* were not entirely welcome when initially published. Even today, there are a surprising number of skeptics. His concepts have survived only because they are much more consistent with modern scientific observations, including our knowledge of DNA and the role it plays in biological reproduction.

One would expect the prefrontal lobe to play a key role in the creative process. The prefrontal lobe plays a central role in our day-to-day experience. It integrates our sensory and emotional experiences. It is uniquely responsive to our dopamine reward system and oversees our behaviors.

We are beginning to see evidence confirming the prefrontal lobe's role in the creative process. The medial (more centrally located) prefrontal cortex is closely related to the subcortical nuclei of the limbic area, which includes the dopamine reward system. It also appears that the medial prefrontal cortex is the first area of the prefrontal lobe to become functional.

Researchers have discovered that when jazz musicians

improvise, the medial prefrontal cortex is consistently active.[38] They evaluated the brain activity of jazz pianists using a functional magnetic resonance imaging (fMRI), which measures the activity of specific brain areas. The medial prefrontal cortex was consistently active during jazz improvisation. In contrast, the dorsal lateral prefrontal cortex showed decreased activity. Dorsal lateral refers to those areas of the prefrontal cortex that are positioned above and lateral to the medial prefrontal cortex.

The medial prefrontal cortex is involved in self-expression while the dorsolateral prefrontal cortex is responsible for self-censoring and inhibition. The medial prefrontal cortex is operational well before the dorsolateral prefrontal cortex. For evidence of this process in action, you don't have to look any further than your average two-year-old and contrast his or her behavior with a fifteen-year-old and a thirty-year-old. Notice that the two-year-old has no trouble with self-expression. The maturation process involves putting restraints on self-expression. This is a work in progress in fifteen-year-olds. I'm not entirely certain that it's completed in many thirty-year-olds.

The medial prefrontal cortex also plays a prominent role in our sense of self. The medial prefrontal cortex is among the brain regions with the most active baseline activity at rest.[39] I suspect that meditative techniques enhance medial prefrontal cortical activity by reducing the self-censoring and inhibitive activity of the dorsolateral prefrontal cortex. This may account for a sense of being centered during meditative activity and a feeling of being at one with the world.

A decrease in self-censoring and inhibitive activity of the dorsolateral prefrontal cortex may free the medial prefrontal cortex to explore more freely new ways to connect the dots. I suspect that this is why artists indicate a need to be "in the mood" for creative activity and athletes report functioning best when they are "in the zone."

Creativity involves a balance between fantasy and perceived

reality. Dreaming, for example, involves a suspension of prefrontal control of neural network associations. Nonetheless, the fanciful associations that occur in dreams can be helpful. The discoverer of the ring structure of benzene (the German chemist, August Kekule, in the mid-1800s) reported that the idea came to him in a dream in which he saw a snake bite its own tail to form a ring.[40] This insight into the configuration of benzene molecules (carbon atoms arranged in a ring pattern) is at the foundation of modern biochemistry.

INTERACTIVE CONNECTIVITY

Interactive connectivity refers to a condition in which elements of a group are in a state of interaction. The term is used primarily with computer systems. In a computer network, the elements of the system are connected in such a way that they are able to influence and be influenced by one another.

Every complex system can be thought of as a network of interactive connectivity. The complex systems are as diverse as the prevailing wind patterns in the earth's atmosphere and the life-forms found in the earth's biosphere. Interactive connectivity is particularly obvious in social systems, whether they be family, tribe, nation, or a group of nations.

The brain is a network of interactive neural connectivity. The different areas of the brain are continually interacting with one another, forming a dynamic neural network. Areas of the brain that are involved in sensory perception and behavior are interactively connected to one another.

Interactive connectivity permits the creation of new and different patterns of association. Some forms of interaction are strengthened and endure while others are weakened over time. This process underlies the evolution of social systems, economic systems, and brain function.

Brain interactivity permits pattern of sensation and behavior to be closely associated in a manner that promotes survival. Human survival is closely associated with social interaction. As you would expect, close connectivity is maintained between patterns of sensation and behavior that promote social welfare. We are very responsive to behavioral cues, verbal and nonverbal. Language skills, in particular, enable humans to attain a high degree of socially coordinated activity.

The brain's network of interactive connectivity creates a reality that is inherently oversimplified. Our reality is a pattern of connectivity developed by the brain to facilitate survival and social existence. It reduces the complexities of our environment to manageable patterns that enable us to maintain a sense of security and satisfaction.

By forming original patterns of association, the brain is the source of human creativity. The brain is continually creating imaginative patterns of association as we process our experience. Often these associations are fanciful and don't resemble anything that exists outside of our imagination, such as flying carpets and genies in a bottle. However, the same creative process underlies the advancement of modern science.

D. IMPORTANCE OF AFFINITY

Emotional reactions provide a critical dimension to human relations. Vision, hearing, and touch are purely descriptive. In addition to seeing and hearing you, my emotional reactions let me know with a high degree of accuracy whether I should welcome or avoid you.

Emotional reactions govern our attraction for one another. Attraction may range from a high degree of attraction to an absolute absence of attraction, possibly even repulsion.

By affinity, I'm referring to an *intermediate degree of attraction*.

Complex systems require an intermediate degree of attraction. Affinity (an intermittent degree of attraction) permits the components of a complex system to remain interactive for extended periods. A high degree of attraction markedly limits interactivity. The component parts settle into a static organizational state. The absence of attraction results in an absence of organized activity patterns (a state of chaos).

Affinity underlies social organizations. High affinity tends to characterize family and tribal relationships. A lesser degree of affinity tends to characterize the relationships between members of larger social networks, such as a nation state. An absence of affinity characterizes the relationship of individuals that are indifferent to one another. The opposite of attraction is associated with conflict.

Affinity permits the elements of a complex system to influence and be influenced by one another. In such a setting, it is possible for new forms of connectivity to evolve over time. Earth's gravity, for example, provides the intermediate level of attraction needed for water and air molecules to be sufficiently interactive to form the earth's biosphere.

Communication plays an important role in social affinity. Shared language contributes to the formation of nation states. Advances in communication technology are associated with social transformation. The written records have been a source of stability and cultural advancement since the earliest historical period. History has been shaped by increasingly wider and more rapid distribution of information. Advances in electronic interactivity, such as Internet connectivity, are ushering in an even newer phase of global interactivity.

Monetary systems also play a prominent role in promoting social interactivity. Monetary interaction permits buyer and seller to exchange goods despite less than perfect social relationships. Monetary relationships often lead to improved social relationships.

TRANSACTIONAL AND NONTRANSACTIONAL RELATIONSHIPS

While monetary systems and communication technology facilitate social interaction, interpersonal relationships are critical to human existence. The family has served as the key to the socialization of every new generation. Our sense of self is founded upon our interactions with others. The importance of personal relationships extends throughout our lifetimes.

It's useful to distinguish between transactional and nontransactional relationships. Both types of relationship involve an intermittent level of connectivity that I refer to as affinity. However, nontransactional relationships are uniquely important for human beings.

Transactional relationships involve an exchange. Our economic system, for example, is based on transactional relationships. Currency transfer enables a society to maintain an intermediate level of social connectivity in which a variety of new industries and markets can evolve.

The ability to reliably engage in transactional relationships is referred to as competence. Competence develops through transactional experience with positive outcomes. A competent businessman is one who has engaged in successful business transactions. Individuals without business experience who believe that they're born businessmen are usually deluding themselves.

Nontransactional relationships, on the other hand, reflect a valuing of the relationship independent of transactional factors. Family relationships are often nontransactional. A marital relationship, in which each partner values the other without a strict quid pro quo accounting, is a form of nontransactional relationship. Our relationship with our children is often nontransactional.

Transactional and nontransactional relationships do not exist in isolation. Both types of relationship tend to coexist. The nontransactional relationship between a husband and wife is strengthened by their transactional relationship. Likewise, the transactional relationship between business leaders is strengthened by their nontransactional relationship—their genuine respect for one another.

Both transactional and nontransactional relationships are forms of affinity. Both contribute to the maintenance of an intermediate degree of interactivity (between order and chaos) in which social and economic evolution can take place.

Both transactional and nontransactional relationships promote security. Security is a sense of diminished anxiety in the face of life's uncertainties. A sense of competence enables an experienced individual to face new challenges in a positive manner. A competent person is somewhat immunized from fear of failure.

A sense of security based upon nontransactional relationships is particularly important. A successful marriage is dependent upon the couple's nontransactional relationship. Do they value each other and enjoy one another's company? In my experience as a psychiatrist, the marriages that could not be saved were those in which the couple had lost touch with the nontransactional aspects of their relationship and had only stayed together for the children's sake. The relationship with their children, their only remaining nontransactional relationship, was all that kept them together. Once the children left, there was nothing to hold them together.

A sense of security based upon the nontransactional relationship with one's parents is particularly important for young children. The stability of their developing sense of self is based upon a deep conviction of their worth. Children that are traumatized by their parents or by the loss of their parents find it more difficult to gain a sense of security. They often go through life trying to compensate for residual insecurity.

The importance of nontransactional relationships is also evident in our major religions. Christianity teaches us to love our neighbor as ourselves. Every major religion stresses the importance of nontransactional relationships. It's becoming increasingly obvious that our ability to survive as a species depends upon a recognition of the value of all human beings.

I-THOU RELATIONSHIPS

Martin Buber, a Jewish philosopher and theologian (1878–1965), made a similar distinction in his book *I and Thou*. Buber distinguishes between I-It and I-Thou relationships.[41] Buber focuses on the quality of relationships.

I-It refers to relationships between a person and things. I experience the object, analyze it, and use it in pursuit of my interests. This is our most common manner of relating to objects. Buber emphasized that we often treat people in the same manner. Economic theory assumes that interactions with one another are on an I-It basis.

An I-Thou relationship refers to a reciprocal relationship that is possible between human beings. I'm able to interact with another human being in a much more emotionally meaningful manner than is possible with an object. I-Thou relationships involve us at multiple levels—cognitive and emotional.

I-Thou relationships are transformative. I-Thou relationships challenge our current sense of self and nurture growth. I-Thou relationships promote a level of emotional security that permits experimentation. A one-year-old will actively explore his or her surroundings, provided the mother is nearby and available as needed. Within the scope of I-Thou relationships, we have the freedom and safety to develop a stronger, more competence-based sense of self.

E. UNCONSCIOUS BRAIN FUNCTION

Unconscious factors permeate our thinking and behavior. We bring to every interaction a set of emotional biases. A person's emotional biases can result in endlessly repeated self-defeating behavior. In therapy, I focus my attention on the emotional aspects of behavior. How did you feel when that occurred? What did you hope would happen? A person's cognitive understanding of his or her own behavior is often a rationalization rather than an explanation. The better we understand our behavior, the more likely we are to modify it in response to current circumstances.

Our brains are specifically adapted to deal with human social behavior. We're able to automatically make useful inferences regarding another person's behavior. We're able, for example, to quickly infer if our behavior has pleased or upset them. If we've been misunderstood, we can quickly correct the situation. Our adaptation to deal with social behavior biases us to the same types of inferences in response to other types of events. We have a long history, for example, of identifying natural phenomena with actions of gods that we can influence by our prayers and behavior (sacrifice and good deeds).

We also have a strong need for certainty and closure. We automatically reduce complex phenomena to the limited number of factors we're able to process. Once we've arrived at an understanding of reality in those limited terms, we cling to it tenaciously. I'm reminded of our legislative body's approach to complex social issues. Each political party has strong views regarding the righteousness of their beliefs and the misguided efforts of their opposition. Religious groups routinely maintain a sense of security by embracing their beliefs as absolute truth despite the uncertainties of day-to-day existence and the wide range of conflicting beliefs held by other religious groups.

Psychiatry has a long tradition of interest in the unconscious.

Most schools of psychiatry are based upon the assumption that unconscious influences are pervasive in human behavior. Carl Jung (1875-1961), an influential Swiss psychiatrist, described consciousness as "like a surface or a skin upon a vast unconscious area of unknown extent."[42] His goal in therapy was to work toward a balance between the conscious and the unconscious.

JUNGIAN PERSPECTIVE

Carl Jung presented an overview of this conceptual framework in a series of lectures in 1935 that were later published in book form titled *Analytical Psychology: Its Theory and Practice.* His observations provide me with an opportunity for further clarification of the relationship between the unconscious and the psychosocial aspects of brain function.

Dr. Jung points out that the conscious mind "can hold only a few simultaneous contents at a given moment." I stress our inability to deal with the full complexity of our environment.[43] We automatically reduce the complex phenomena to a limited number of parameters that our brains are able to process.

I've indicated that consciousness is the product of the brain's cortical-cortical connections under the influence of the prefrontal cortex. Jung points out that "nothing can be conscious without an ego to which it refers. If something is not related to the ego then it is not conscious."[44] "The ego is a complex datum which is constituted first of all by a general awareness of our body, of your existence, and secondly by your memory data; you have a certain idea of having been, a long series of memories."[45] I indicate that the brain's cortical-cortical connections are under the guidance of the prefrontal cortex, and our sense of self is constructed by connecting the dots of our experience.

Jung indicates that the ego "has the power of attraction, like a magnet; it attracts contents from the unconscious, from the dark realm of which we know nothing; it also attracts impressions from the outside, and when they enter into association with the ego they are conscious."[46] I indicate that the self is a complex system and, as such, is a "strange attractor" that stabilizes our identity while being subject to discontinuities as seen in other complex systems (more about discontinuities in the following section, which addresses trauma). Strange attractors and system discontinuities will be explained as principles of complex theory in part IV of this book.

Jung distinguishes between the personal and collective unconscious. The personal unconscious is based upon personal experiences that influence our behavior.[47] Our emotional memories develop independently from our conscious (declarative) memories. An individual whose hippocampus has been removed won't remember talking to me shortly after I leave the room. However, when I return, despite having no conscious memory of our previous meeting, he may experience fear if I kicked him before I left.

Jung also emphasizes the collective unconscious. The collective unconscious refers to unconscious patterns that are shared by mankind. He refers to such patterns as archetypes. Archetypes are reflected by well-known motifs, such as the hero, the redeemer, the dragon. These motifs are found in myths, legends, and fairy tales reflecting the deepest regions of our unconscious.[48] I emphasize the pervasive influence of our subcortical systems involved in approach/avoidance and the maintenance of interpersonal connectivity (affinity). These emotional predispositions are the products of our evolutionary pasts.

TRAUMA

Trauma is an assault on our sense of self. We all have negative experiences. Fortunately, we're not overwhelmed by day-to-day negative experiences. Trauma refers to negative experiences that are overwhelming.

Trauma tends to produce dissociation—a failure to incorporate traumatic experiences into our sense of conscious awareness. In post-traumatic stress disorder, for example, traumatic past experiences often intrude on current day-to-day experience. Current events that would otherwise be unthreatening can touch upon this area of dissociated experience, causing it to flood back into consciousness as if it were reoccurring (referred to as a flashback).

Dissociation can be extremely severe. While a traumatized adult may develop post-traumatic stress disorder, a young child may develop even more severe forms of dissociation, such as dissociative identity disorder (more commonly referred to as multiple personality disorder). In this disorder, traumatic experiences are associated with alters—personality fragments formed in response to traumatic experiences. A person with dissociative identity disorder goes through life with these personality fragments coming out unpredictably in response to current situations.

OVERLEARNED BEHAVIOR

Overlearned behavior refers to activity that has been repeated so often that it can be performed without reliance upon conscious prefrontal processing. Prefrontal conscious processing slows down our reactions. It's often important that we react more rapidly than prefrontal conscious processing can support.

Martial arts is a form of overlearned behavior in response to attack. The actions involved are performed very rapidly without being slowed by the time required for the prefrontal lobe to fully process what's taking place.

Typing (as performed by a skilled typist) is another form of overlearned behavior. As I've put this book together, it's become increasingly obvious that I'm not a skilled typist. I have to fumble around for the next key and spend a good portion of my time correcting mistaken entries. A skilled typist, on the other hand, can blaze through the material at breathtaking speed (breathtaking to me, at any rate) with minimal errors. The motions of a skilled typist are overlearned and can be performed without consciously focusing on individual keys.

Speech is a particularly important form of overlearned behavior. The neural network associations that process language are so highly overlearned that you and I can perform this activity with only a fraction of conscious attention. I don't have to strain to understand or express myself in English. I can begin a response even as I'm weighing the turn of phrase that will best convey my meaning. My facility with English language is in sharp contrast to my lack of facility in any other language.

Highly learned behavior depends upon well-established neural network connections that are able to function automatically with only a fraction of conscious attention. As every pianist knows, it is the result of endless repetition.

F. MALE-FEMALE DIFFERENCES

Differences in brain function and information processing between males and females are to be expected. After all, the developing brains of young males and females are exposed to different hormonal environments. One obvious difference is in testosterone exposure during fetal development. In general, boys have more exposure to testosterone during the fetal period. Testosterone is an anabolic steroid. Anabolic steroids act upon every organ system in the body, including the developing brain.[49]

Sexual differences offer a unique opportunity for insight into brain functions. Much of our understanding of brain function is based on brain injury. Damage to a particular area of the brain, for example, is associated with specific abnormalities in brain function. Rather than focusing exclusively on brain injury (Where's the lesion?), we're able to gain useful information concerning brain function by examining differences in cognitive processing between the normal brains of males and females.

I'll focus upon one specific aspect of brain function—brain lateralization. Lateralization refers to the degree in which the brain compartmentalizes function to either the right or left hemisphere. Lateralization reflects the degree to which neural activation differentially involves either the right or left hemisphere (as opposed to being generalized in both hemispheres).

Lateralization influences the way information is chunked. In order to deal with the vast volume of data we encounter, our brains learn to combine it into more manageable chunks of information. Differences in male-female lateralization would result in differences in information chunking.

A growing amount of evidence indicates that, in general, men and women differ in regard to brain lateralization.[50] Please note the qualifier: "in general." I'm referring to a statistically significant difference rather than an absolute difference. There are plenty of exceptions and a good deal of overlap. Nonetheless, I find that an awareness of male/female brain lateralization differences is very useful in counseling heterosexual couples. Heterosexual couples are more likely to encounter difficulties arising from differences in male/female brain lateralization.

Men tend to lateralize brain function more than women. When men solve spatial problems, for example, there tends to be increased activity involving the right hemisphere, which specializes in spatial processing. When men are asked to solve verbal problems, there tends to be increased activity of the left hemisphere, which specializes in verbal processing.

On the other hand, when women are asked to solve either type of problem, there tends to be symmetrical activity over both hemispheres. This observation was initially based upon several simple studies using electroencephalographic measures of brain activity. Studies using more sophisticated functional brain scans have confirmed the observation.

It is suspected that women have more connectivity between the right and left hemisphere than men. There is some evidence, for example, that the posterior portion of the corpus callosum, the primary neural connection between the right and left hemisphere, is larger in women. The posterior portion of the corpus callosum connects the sensory cortex of the right and left hemispheres. Interconnection between the right and left hemisphere may also be influenced by dendritic density, which isn't as easily measured.

Men tend to chunk information in a unidimensional manner, reflecting their greater reliance upon lateralization (also referred to as compartmentalization). Men tend to organize information in terms of either visual-spatial symbols or verbal symbols—but not both at the same time. Men gravitate toward visual-spatial symbols in their recreational activities (e.g., watching football, a particularly visual-spatial pastime) and occupational activities (e.g., the military relies heavily upon visual-spatial symbols, such as uniforms, flags, and organization diagrams).

Women tend to chunk information multidimensionally, reflecting their reliance upon information from both hemispheres at the same time. They chunk information in a manner that integrates the full range of the sensory information available to them. I can only think of one way to usefully chunk the visual-spatial information associated with right hemisphere function and verbal information associated with left hemisphere function— in terms of specific people (not mankind in the abstract) and the relationships between people. Recreational activities (e.g., soap operas) and occupational activities (service industries) commonly associated with women tend to stress the importance of interpersonal relationships. Women tend to be much more aware of relational issues than their male spouses are.

Men tend to organize information in terms of visual-spatial maps. They feel more secure after orienting themselves with a map. Visiting an art museum, the husband begins by seeking out a museum map. His wife, on the other hand, is focused on observing pieces of art. Unfortunately, he has a tendency to interpret her activity as impulsive and disorganized while she concludes that he is a hopeless stick-in-the-mud who is more interested in the museum map than the surrounding art.

Men are also likely to organize information in terms of verbal symbols. The fellows place a lot of emotional investment in verbal rules. Men are the ones who come up with phrases such as, "A man's got to do what a man's got to do" while finding it difficult to appreciate their wife's argument that a man's highest priority should be to take care of his wife and children.

I believe it is more or less accepted that many of the world's major religions have a particularly strong male influence. Who else would put such a strong emphasis upon the power of the "Word?" Women are much more focused upon relationships. In that regard, it appears that women are more in touch with the essence of religion than men are. When Christ was asked which is the most important commandment, he responded, "Love God with your whole heart and soul; and love your neighbor as yourself." He focused upon one's relationship with God and with one's fellow human beings rather than on specific rules. Every major religion emphasizes the importance of human relationships.

Women tend to organize information in a less obvious manner. How do you organize information that is chunked in terms of people with emphasis upon relationships between people? I can only think of one way—sequentially. Directions tend to be provided in terms of a series of steps. This is a surprisingly effective way to remember many things. It's clearly useful in remembering anniversaries and birthdays that serve as sequential links in the chain of memories that encompass a lifetime.

A sequential approach often enables women to deal routinely

with a higher level of complexity than the fellows. Interpersonal relationships are highly complex and can easily extend beyond the comfort level of the guys. Men may feel downright uncomfortable watching a soap opera. The complex relational matrix of a soap opera is not easily reduced to a visual map or a few verbal rules. Men are often oblivious to the nuances of children's behavior. Their wives can't understand how they fail to recognize the "obvious."

I've noticed a difference in something as mundane as grocery shopping. I can't remember where most items are located in a large grocery store. My wife, on the other hand, can give me a list that follows the items from one aisle to the next. Her sequential organization of information enables her to remember where most of the items are located.

The differences in information processing between men and women easily result in misunderstandings and conflict in a marriage. Marital conflict creates insecurity for both husband and wife. Insecurity, in turn, tends to magnify male-female differences. On one occasion, both the husband and the wife complained that they were handling 75 percent of the marital responsibility. They used essentially the same words to describe their frustration in the marriage. Each was simply reporting from his or her own perspective. They were referring to different aspects of the relationship. A marital counselor often functions as a translator.

Marital counseling involves helping the couple to understand and value each other. Men and women have different strengths and weaknesses. In many ways, they complement one another. However, their differences can be valued or devalued. Usually, couples value their differences early in a marriage. Those requiring marital assistance have a strong tendency to devalue their differences. As they understand themselves and each other better, it is possible to appreciate how they complete one another and to value their differences again.

Please keep in mind that differences between men and women in regard to brain lateralization are not black and white. I'm referring to a phenomenon that spans a spectrum of possible lateralization differences. In my experience, an awareness of differences in information chunking between men and women is most useful when I'm working with heterosexual couples. With heterosexual couples, there tends to be relatively predictable differences between complementary patterns of organizing information. I have to be cautious about drawing upon these observations when I'm working with homosexual couples.

G. PSYCHOTHERAPY

Adaptation involves a feedback process. We observe, assess, and act in a recurring cycle. Therapy attempts to intervene in this process.

Behavioral, cognitive, and dynamic psychotherapy attempt to influence one or more phases of this cyclic process. Behavioral therapy focuses upon the action phase. It's hoped that changed behavior will impact all aspects of the cyclic process. Cognitive therapy focuses upon the observation/assessment phase. It's assumed that the individual is failing to accurately perceive the consequences of behavior. In addition to observation/assessment and behavior, dynamic psychotherapy focuses upon the unconscious aspects of behavior. Unconscious factors influence every aspect of behavior. Bringing the unconscious into consciousness can help modify disruptive behavior.

Supportive therapy is the most widely used form of therapy. Supportive therapy focuses on improving our day-to-day sense of satisfaction. How could the patient handle a problem differently to obtain a more satisfactory outcome? Does disappointment in one situation mean that failure is inevitable in every situation? Our sense of security is directly related to our sense of confidence that we can deal with life's challenges successfully. Success breeds confidence. Repeated failure is associated with increasing depression.

Over time, all of these approaches impact the actual functioning of the brain. The malleability of the brain and the impact of experience upon brain function underlie human adaptation and psychotherapy.

During the past sixty years, a growing number of psychoactive medications have been developed that impact brain function, often more effectively than talk therapy alone. Many of these medications influence a particular neurotransmitter system. Antidepressant medications appear to interact with serotonin and norepinephrine systems. Antianxiety agents interact with GABA receptors. Antipsychotic medications interact with dopamine receptors.

Overall, our medications work by influencing patterns of neural oscillation. These medications influence neural firing rates, which, in turn, influence oscillation rates as the brain responds to stimuli. You may be surprised to learn that a number of medications used for psychiatric disorders were initially introduced as antiseizure medications. These medications raise firing thresholds of neurons and modify the pattern of neural oscillation in that manner.

Electroconvulsive treatment also influences neural oscillation rates. Electroconvulsive treatment may work in somewhat the same manner as cardioversion. In cardioversion, abnormal cardiac activity is disrupted by an electrical stimulus. As cardiac activity returns, the heart is able to resume a normal beating pattern. In some cases, electroconvulsive treatment is recommended because improvement tends to be more rapid and the patient experiences fewer adverse side effects than are associated with long-term use of medication.

In recent years, transcranial magnetic stimulation has become increasingly utilized in psychotherapy. Transcranial magnetic stimulation involves placing a small magnetic field generator near the patient's head that generates repetitive pulses of magnetism. This form of therapy directly alters oscillation patterns using repetitive magnetic pulsing. Location of the magnetic field generator and pulsing frequency are important variables in this form of treatment.

All forms of therapy influence actual brain function. Over time, therapy influences the pattern of neural oscillation. Safer, more rapid forms of therapy are constantly being sought. Our understanding of brain function is expanding rapidly. It's an interesting time to be involved in psychiatry and neuroscience.

PART IV ———————————————————

COMPLEX SYSTEMS: GUIDING PRINCIPLES UNDERLYING BRAIN FUNCTION

A. SETTING THE STAGE

Matter and energy are interchangeable: $E = mc^2$. A mass of weapons-grade uranium may sit relatively unchanged for years—a form of potential energy. When brought to a critical mass, there is a massive release of kinetic energy. Order has been transformed into chaos.

Between the tight bonding of solid matter and the extreme disruption of bonding in a nuclear explosion, there is an intermediate level of bonding. Water, for example, refers to a liquid phase between ice and water vapor. In the liquid phase, water molecules are interactive but not locked into fixed positions. The same is true of the air molecules of the earth's lower atmosphere. The air molecules that we breathe are held in close proximity to one another by the earth's gravity. Like water molecules, they are fluidly interactive without being locked into fixed relationships with one another.

The intermediate bonding of molecules of the earth's oceans and lower atmosphere serve as a holding environment in which complex systems (which includes living systems) can evolve. In this unique environment, the components of complex systems are able to interact with one another. *Complex systems are sustained patterns of interaction* that evolve in such a holding environment. Between order and chaos, one finds complex systems.

In addition to a holding environment, complex systems require a reliable energy source. A holding environment holds the components of a complex system within an interactive relationship. Energy enables the components to remain interactive and to develop enduring patterns of interaction.

Our sun provides the energy needed for evolution of complex systems (including living organisms) in the earth's biosphere. All complex systems are energy dependent. Life on earth cannot be sustained without a reliable, ongoing source of energy.

Complex systems are not eternal. Every life-form, for example, has a limited lifespan. Earth's life-forms have countered this limitation through a reproduction process that enables life-forms to pass on from one generation to the next their genetic plan. However, species come and go despite their reproductive capabilities.

Repetitive activity is characteristic of complex systems. Our solar system, for example, is defined by the repetitive orbiting of the planets around the sun. Every sustained process of complex systems involves a form of repetitive activity. Our respiratory and circulatory systems, for example, depend upon the repetitive activity of breathing and the heart's pumping action. Oxygen-carrying blood is constantly being pumped from the heart to the body, making oxygen available throughout the body and carrying away carbon dioxide (a waste product of metabolism). The blood then returns to the heart, taking on oxygen and disposing of carbon dioxide as it passes through the lungs and repeats the cycle.

In this section, I'll review the general principles that underlie complex systems. These principles apply to brain development and function. My confidence in the importance of neural network oscillation in brain processing is based upon the role of repetition in complex systems.

Before we begin our discussion of complex systems, I'd like to emphasize two additional points. The first point is that inhibitory factors play an important role in complex systems. We're able to utilize uranium for energy production by inhibiting the speed of nuclear fission through the use of control rods that absorb neutrons produced by uranium fission. By adjusting the control rods, reactor operators are able to adjust the intensity of nuclear activity. The control rods enable us to sustain the process over time and use it for energy production (rather than mass destruction). The brain relies upon inhibitory interneurons to maintain neural network oscillatory activity within the necessary parameters. The speed of oscillation is controlled by varying the inhibitory field in the cortical columns.

Repetitive patterns of activity involve a balance between the forces that stimulate and those that inhibit activity. The population of a particular animal species, such as rabbits, is maintained by a balance of available food, reproductive rate, life span, and predators. Over time, a relatively stable rabbit population evolves if the balancing forces remain unchanged. Modification of the forces at work, such as a reduction in the number of the rabbit's natural predators, is associated with a change in the rabbit population.

The second point that I'd like to emphasize is that complex systems are not linear phenomena. Every complex system represents a balance of forces. A hurricane, for example, is the product of the earth's rotation, the temperature of the tropical ocean, and the rising warm-water-saturated air from the ocean surface. The multiple factors involved in creating and sustaining a hurricane don't lend themselves to linear analysis.

We rely strongly on linear explanations. Our tendency is to conclude that if c follows b, and b follows a, then a accounts for *b* and *c*. Linear explanations are very useful in many situations. They enable us to handle many everyday events rapidly and effectively.

However, linear explanations can easily lead to false conclusions when applied to complex phenomena. The world's diverse religious traditions, for example, tend to rely upon linear explanations to account for our existence. These explanations typically involve a god or gods who, through a linear series of actions, have created mankind and the world around us. We tend to deal with a god or gods by acts of appeasement and sacrifice. It is becoming increasingly clear that the evolution and sustainment of life on earth is far more complex than these linear explanations are able to convey.

B. SAMPLE COMPLEX SYSTEM: HURRICANE KATRINA

In late August 2005, tropical storm Katrina grew to a category 1 hurricane and made landfall in southern Florida with 80 mph wind and trailing heavy rain. It crossed Florida and quickly grew to a category 5 hurricane in the warm waters of the Gulf of Mexico, with winds topping 175 mph. It made landfall in southeastern Louisiana as a category 3 hurricane with 125 mph winds. As it crossed land, reaching Tennessee, it was downgraded to a tropical depression. Remnants of the storm extended all the way to eastern Canada.[51]

The satellite photograph of a hurricane (figure 5) illustrates several characteristics of complex systems. The atmospheric components of the hurricane are repetitively circulating around the eye. A hurricane is defined by wind speed and its cyclic pattern of atmospheric activity. The energy of air rising from the warm waters of the Gulf of Mexico is channeled back into the storm system rather than being allowed to dissipate—a form of feedback—raising wind speed to hurricane strength. Feedback is required to maintain every complex system.

FIGURE 5: SATELLITE PHOTOGRAPH OF A HURRICANE

A hurricane represents a discontinuity in the earth's global weather patterns. It is a disruption (discontinuity) in the normal patterns of our global weather system. Discontinuities are characteristic of complex systems.

Earth's gravity supports an envelope of air and water that provides sufficient connectivity for the evolution of complex systems. Our atmospheric and oceanographic flow patterns are complex systems sustained by the earth's holding environment. Even discontinuities, such as a hurricane, are a form of complex system that develops at the unstable extremes of the more routine air and water flow patterns.

We'll review the following basic concepts: complex systems, weak/strong strange attractors, emergent phenomena, open systems, usable energy, energy gradients, steady states, feedback loops, holding environments, the logistics curve, system discontinuities, and incremental change through jugglery. We'll also review the role of information transfer involving complex systems. Information transfer underlies biological evolution, social evolution, and personal development.

C. CHARACTERISTICS OF COMPLEX SYSTEMS

Complex systems involve the interaction of three or more interactive elements. Interaction involves an intermediate level of bonding between the interactive elements. As I indicated earlier, an intermediate level of bonding underlies the interactivity of water molecules and air molecules of the earth's atmosphere.

Complex systems are sustained patterns of interaction that evolve among interactive elements in a setting of intermediate level bonding. A hurricane is recognized by the organizational pattern of the components that make it up. Air molecules are constantly feeding into and separating from the structure of a hurricane. The same is true for other types of complex systems. Social organizations also involve an intermediate level of bonding that enables them to persist while individual society members come and go.

Complex systems are sets of interactive elements that exhibit emergent properties. Emergent properties are not obvious from the properties of the individual elements. This isn't as complicated as it sounds. Metropolitan traffic systems, for example, are the product of huge numbers of cars moving back and forth from home to various locations on a daily basis. Patterns develop over time. Those patterns can't be predicted by studying an individual car. Likewise, the earth's wind patterns and ocean currents are products of large-scale molecular interaction that can't be predicted by examining individual air or water molecules.

Because complex systems involve at least three (usually more) interactive elements, their pattern of interaction is difficult to predict in a precise manner. Systems that can be understood in terms of only two interactive elements tend to be highly predictable. The orbit of Earth, for example, is dependent almost entirely upon interaction between Earth and the sun. As a result, we're able to make very precise predictions regarding Earth's location one hundred years from now. Earth's weather, on the other hand, involves the interaction of the vast number of elements in the earth's atmosphere—air molecules, humidity, temperature, rotation of the earth, terrain features, etc. We're unable to predict weather with precision from one day to the next in many areas. Likewise, we're unable to predict human behavior with precision.

All complex systems involve recurrent activity. The air molecules of a hurricane recurrently circle the center of the storm. Without recurrence of activity, a complex system simply dissipates and cannot be sustained. Complex systems are identified by their pattern of motion.

WEAK/STRONG STRANGE ATTRACTORS

Due to their unpredictability, complex systems may be thought of as "strange attractors."[52] They are attractors in that they foster a sustained pattern of activity. The pattern of attraction in complex systems is described as strange because the ongoing pattern of activity is not precisely predictable.

Strange attractors span a range of predictability from weak to strong. The more predictable types are strong attractors while the less predictable are weak attractors. A hurricane is a relatively strong strange attractor compared to a tropical storm. A hurricane has a much more clearly defined pattern of dynamic activity that is more sustainable than a tropical storm.

The evolution of living organisms has involved the development of increasingly more reliable (strong) attractor systems. Living organisms require sustained, highly reliable interaction among their internal components and with their surroundings.

The human body, for example, requires a strong attractor system. Without a regular heartbeat and a constant supply of oxygen, we can't survive. Our temperature must be maintained within narrow parameters. Even a brief lapse in body function can result in death.

Brain development and function requires an exceptionally strong attractor system. Seizures and loss of consciousness (disruptions of the brain's attractor system) are potentially life threatening. Less obvious aberrations in brain function can have equally serious consequences. Individuals with schizophrenia or mania, for example, may behave in a bizarre fashion due to erratic brain function and a distorted sense of reality.

EMERGENT PHENOMENA

Brain function is an emergent phenomenon. An emergent phenomenon is a synthesis in which the whole is greater than the sum of the parts. The symphonic performance of an orchestra, for example, is an emergent phenomenon with the sound of each instrument contributing to the overall performance. The overall pattern is more than the sum of the notes played by the different instruments. It is distinguished from "noise" by the timing and coordination of the sounds involved.

While we know a lot about the action of specific neurons, we are only beginning to understand the emergent activity of neural networks. Brain function, as is true of every complex system, is an emergent phenomenon. It springs de novo from the motion and relational patterns of its component parts.

Sound is a familiar example of an emergent phenomenon. Sound waves result from air molecules vibrating en masse. The vibration of air molecules is passed to surrounding molecules and spreads—similar to the wave produced by a rock falling into a pond. In a vacuum, there are no sound waves. Even in a partial vacuum, molecules lack the proximity needed to propagate sound waves. Sound waves are emergent phenomena that require large numbers of interactive air molecules. They can't be explained by studying individual air molecules.

Emergent phenomena characterize every phase of existence on planet Earth. Complex phenomena evolve in an emergent fashion from preexisting components. The atomic elements allow for the creation of endless numbers of compounds and chemical reactions. Living organisms have evolved from the interaction of these compounds. We're unable to understand living organisms by studying the compounds that make them up, just as we cannot appreciate a symphony by studying individual notes.

Brain function and development is an emergent phenomenon that takes place over the course of a lifetime. Early phases of neural axon development are the product of chemical interactivity. This phase of neural development is shaped by chemicals in the general area of a developing neuron. Over time, nerve cells develop axons that bring them into more direct contact with one another.

Once neurons are able to influence one another directly, their interaction is transformed. Groups of neurons are now able to utilize a combination of neural axonal electronic transmission activity and chemical interactivity at the neural synapse. Neural network oscillation is now possible. Over the course of a lifetime, the strength of synaptic connections is shaped by experience. The pattern of neural network activity found in the adult brain is an emergent phenomenon that cannot be explained by simply studying the brain's chemical components.

OPEN SYSTEMS

Open systems are those that gain or lose matter/energy over time. Earth itself is an open system. Earth continually receives energy from the sun and radiates energy into space. Every living organism takes in or loses matter/energy over time.

Entropy refers to the tendency for energy to always move from a higher to a lower energy state. A wooden log, for example, releases energy as is burns. The energy of the log is dispersed into the environment. This is a one-way process. One can think of an energy state as the degree of organization. Organization tends toward disorganization.

A closed system is one in which nothing, including energy, enters or leaves. Closed systems have been critical to our understanding of entropy. In a closed system, it can be demonstrated that reactions take place only when there is a decrease in usable energy. Energy has been dispersed, becoming less organized. Usable energy has been transformed into a form of energy that is no longer as easily utilized.

Every system in the biosphere (the portion of the earth's environment that contains living organisms) takes in and gives off energy. All are open systems. Open systems are inherently more complex than closed systems. More variables are involved, and matter/energy is continually being gained or lost.

USABLE ENERGY

Open systems consume usable energy. Hurricanes, for example, consume enormous atmospheric energy. They dissipate when that energy is no longer available. Species and societies rise and fall depending upon availability of energy and efficiency of energy utilization.

Usable energy is the result of nonuniform energy distribution.

When differences in local energy levels exist, energy tends to move from areas of high energy to areas of less energy toward equilibrium. Energy flowing from areas of higher to those of lower energy is available for work. Electrical energy, for example, can be generated by the energy of falling water. The electric energy generated by a hydroelectric dam may then be diverted for utilization in a variety of ways throughout the surrounding area.

Energy may be either potential or kinetic energy. Potential energy refers to positional energy. The water in a lake has potential energy. The lake water's potential energy is equal to the energy required to raise ocean water from sea level into the lake. Kinetic energy refers to the energy associated with motion. As the water flows downstream from the lake, its potential energy is converted into kinetic energy. Once it returns to the ocean, it no longer has either potential or kinetic energy. Usable energy has diminished.

ENERGY GRADIENTS

We can think of differences in usable energy in terms of energy gradients. Usable energy moves from a higher energy state to a lower one. Niagara Falls, for example, has a huge energy gradient. There is a tremendous difference in usable energy from the top of Niagara Falls to the water at its base. A series of cascades may have the same overall difference in usable energy from top to bottom. However, a series of cascades reduces usable energy through a stepwise series of much smaller, more manageable energy gradients.

Large energy gradients produce energy-driven reactions. As I continuously stack one brick on top of another, the potential energy of the column eventually becomes unstable, and the column collapses. This is an energy-driven reaction. A hurricane is an energy-driven reaction that takes place when the dynamic activity of the earth's atmosphere becomes unstable. It represents a significant disruption of the earth's relatively stable global atmospheric activity.

Entropy-driven reactions, on the other hand, refer to much smaller, more manageable reactions. Instead of waiting for a brick column to become unstable and collapse spontaneously, an entropy-driven reaction might release the potential energy of an individual brick on an as-needed basis to strike a specific target. The energy required to release a single brick is relatively small.

Living organisms have evolved the ability to utilize entropy-driven reactions. Carbohydrates are metabolized utilizing energy gradients that are so small that we have difficulty measuring them. The energy gradients involved consist of the numerous steps in which carbohydrate molecules are progressively broken down into waste products. We can think of this series of steps as an energy cascade.

The use of extremely small energy gradients associated with entropy-driven reactions is essential for living organisms. Energy-driven reactions, such as a hurricane, come and go unpredictably. Small energy gradients enable us to consume energy in a controlled fashion. Imagine the force of Niagara Falls in your body. An energy-driven reaction of that magnitude would be incompatible with life as we know it. The use of small energy gradients enables the body to take maximum advantage of the usable energy available in carbohydrate molecules.

Living systems use enzymes to facilitate reactions across small energy gradients. Enzymes are proteins that facilitate the wide range of chemical reactions required by our bodies. Without enzymes, these reactions occur only to a very limited extent and much too slowly. Enzymes facilitate reactions. The speed of a reaction depends upon the likelihood of the reactants coming together in the correct configuration. An enzyme brings the reactants into a position that maximizes the likelihood of their coming together successfully. Without enzymes, only a small fraction of the reactants would be properly positioned at any given time, and the reactions would take place at much too slow a rate to sustain life.

Enzymes permit living organisms to take advantage of a wide range of possible entropy-driven reactions rather than depending upon uncontrolled energy-driven reactions. Enzymes enable living organisms to explore the possible rather than be limited to the inevitable.

STEADY STATES

Complex systems are a form of steady-state solution to shifting variables. The term "steady-state solution" refers to a relatively stable pattern of activity that evolves over time in response to shifting variables. Our economy, for example, is a form of steady-state solution. It tends to remain relatively stable despite day-to-day changes in financial transactions.

Steady-state solutions reflect a balance when a variety of conflicting forces are in action. The animal population of an island, for example, represents a balance of forces—some of which increase the animal population while others decrease it. The result is a balance or a steady-state solution in which the animal population remains relatively stable, providing the variables that contribute to the balance are not changed significantly.

Steady state is a manifestation of entropy in an open system. Matter and energy are continually gained and lost. The steady state that results represents a balance of gains and losses. Looking again at the plant and animal population of an island, many population mixes are possible. But only a limited number of population mixes are probable. For a population mix to be probable, overall energy requirements need to be balanced with energy availability. The population mixes of higher probability are those that take more efficient advantage of energy availability.

A mountain lake is a form of steady state. The water level

of the lake reflects a balance of water input and output. Despite varying input and output from year to year, the lake tends to remain relatively intact.

A water tower illustrates the utility of a steady-state phenomenon. The water supply in the tower is determined by a balance of water pumped into the tower and water removed for community utilization. As long as the water level is maintained within reasonable parameters, every household in the community has a reliable source of water that is readily available.

Steady-state stability is essential for life. Living organisms require the steady-state stability of the earth's biosphere and the continuous availability of both solar energy and the interactive components from which they've evolved. Social organizations require stable populations. Stable populations require stable food sources. Stable food sources require reliable access to sources of energy, water, and the numerous chemical components they require. The evolution of complex systems depends upon the stability and availability of the various components that contribute to their evolution.

Living organisms take special precautions to ensure that their internal steady state is maintained within safe parameters. The process is referred to as homeostasis. Our bodies, particularly our brains, are absolutely dependent upon the maintenance of internal environments within precise stable parameters. We must maintain stable body temperatures, respiration, nutrition, and body integrity to survive.

Steady states are relative, not absolute. A wide variety of human social organizations are possible. Each represents a steady-state solution to the challenges the group has confronted. They are continually in transition as new variables are introduced. A social organization that works well in one set of circumstances may not work at all in another.

Neural network interaction is another form of steady state. It is an energy-efficient pattern of action that evolves over time in response to our life experience. The pattern of brain function varies from one society to another. This phenomenon is demonstrated by language development. Over time, with repeated exposure and use of your primary language, your brain develops the steady-state organization needed to understand and speak that language. Language fluency is a steady-state solution unique to one's language environment. Likewise, our sense of self and our experience of reality are steady states that the brain has arrived at through interaction with our particular social environment.

FEEDBACK LOOPS

A feedback loop utilizes the output of a system to influence the throughput. In a positive feedback loop, the rich get richer. In a negative feedback loop, the poor get poorer. Living systems utilize both positive and negative feedback loops to maintain homeostasis.

Positive feedback loops sustain a hurricane. The energy of the wind is continually directed around the center of the storm. The circulating air stream reinforces the wind speed of the storm. This positive feedback process progressively increases the storm's wind speed. The storm's velocity is maintained as long as the proper atmospheric conditions are sustained. Positive feedback loops are needed to offset the continual tendency of the storm's energy to dissipate due to entropy.

Ocean currents provide another example of positive feedback loops sustaining a complex system. Specific currents evolve in the ocean basins as a result of solar energy, the rotation of the earth, and the restraining influence of the surrounding continents. In the northern Atlantic Ocean, for example, ocean currents circulate between the equatorial region and the poles between Europe and North America. These currents, in turn, influence global climate.

Evolution is the product of positive and negative feedback loops.

Species that survive are those that maintain an adequate balance in the cycle of reproduction (a positive, life-sustaining process) and death (a negative, life-disruptive process). Throughout this process, individual organisms and species compete for available resources.

Living organisms have evolved particularly complex systems of positive and negative feedback loops for survival. Feedback mechanisms ensure that energy is distributed to those areas of the body where it is needed. As the brain develops, positive feedback loops enable needed neural connections to survive while negative feedback loops result in the elimination of unnecessary neural connections (apoptosis), maximizing brain efficiency.

HOLDING ENVIRONMENTS

Open systems require holding environments. Holding environments are physical environments in which complex forms of interaction may occur and be sustained. Without a holding environment, usable energy simply dissipates. A holding environment brings together the necessary reactants. It also contains or disposes of reaction products. In the holding environment of an oven, the ingredients of a cake are brought together in the quantities at the proper temperature to produce a cake. Although the reactions involved to produce a cake could conceivably take place elsewhere, they are unlikely to occur with any frequency outside of an appropriate holding environment.

Earth provides a holding environment for reactions that are highly unlikely to occur elsewhere in our solar system. Earth's atmosphere and oceans are able to receive and hold onto energy from the sun. Before solar energy is permitted to dissipate into space, it is transformed into other forms of usable energy. The oceans and atmosphere are warmed. In conjunction with the rotation of the earth, this warming effect produces ocean currents and prevailing winds. With evaporation of ocean water, water is carried by the winds over landmasses where it is released forming lakes and rivers.

Earth's holding environment permits the evolution of complex systems. "Complex system" is a general term that refers to a wide variety of sustained patterns of interaction. Prevailing patterns of atmospheric and oceanic flow patterns are complex systems. The term is also applies to both biological and social organizational systems that have evolved over time.

Biological organisms cannot exist without ensuring that their internal biological processes are protected from environmental disruption. The cells, for example, have evolved an impermeable outer membrane that protects internal cellular function. The body is covered with a dermal layer (skin) that keeps bodily fluids from escaping and protects internal structures.

The human brain requires a particularly secure holding environment. Loss of blood pressure, lack of oxygen, loss of nutrient, and abnormal temperature can all result in loss of consciousness and death. The skull and spinal fluid protect the brain from physical violence. Brain cell exposure to drugs, particularly during early development, can be catastrophic.

LOGISTICS CURVE

The logistics curve describes the relationship of population size to the carrying capacity of the environment.[53] The carrying capacity of the environment varies with climate fluctuations, disease, food resources, and so forth. The logistics curve is an S shaped (sigmoidal) curve.[54] On the lower left, the population is extremely small but rising over time. In the middle portion of the curve, there is a relatively predictable relationship between the rising population and the carrying capacity of the environment. This portion of the sloped curve is relatively linear, reflecting a reliable relationship between population size and environmental constraints—food availability, predators, life expectancy, reproduction rate, prevalence of disease, and so forth. The population remains stable as long as it is within the middle portion of the logistics curve. At the upper end of the curve on the upper right, we approach the maximum carrying capacity, and it flattens out over time. Both the lower left and the upper right portions of the logistics curve are subject to unpredictable variations. Unpredictable variations are found in the lower left portion of the curve due to small population size that is easily threatened by unpredictable environmental variables. However, the population curve also becomes unstable at the upper end of the logistics curve. As the population approximates the maximum carrying capacity of the environment, we find points at which the population size may shift unpredictably between two or more possible configurations. We refer to these points as bifurcation points. At bifurcation points, population size may shift unpredictably between two or more possible configurations. It may also become completely chaotic, defying our attempts to even identify a relationship between carrying capacity and population size.[55]

The logistics curve may be used to conceptualize the general relationship between usable energy and any population. Normally it is used to model the population of a species in a particular holding environment, such as the population of an animal or plant species on an island. However, the concept also may be used to understand the relationship between complex systems (including living organisms and neural networks) and the carrying capacity available to them.

The middle portion of the logistics curve is the most predictable and conducive to the stability required by living organisms. Living organisms use homeostatic mechanisms to proactively maintain their biological systems in the middle portion of the logistics curve. They use enzyme systems to ensure cell function in that specific portion of the logistics curve.

As we approach the upper limit of the curve (the maximum carrying capacity), the regularity of the logistics curve breaks down. The curve reaches a bifurcation point where the population may move toward different outcomes. Earlier, I used the example of a brick column constructed by repeatedly placing one brick on top of another. A bifurcation point is reached when adding one more brick is equally likely to raise the column or cause the column to collapse. It could go either way. This point is referred to as a system discontinuity. As we move further to the right, the logistics curve may encounter multiple discontinuities, even becoming chaotic (completely unpredictable).

SYSTEM DISCONTINUITIES

System discontinuities are points on the logistics curve in which minor phenomena may result in major changes in the evolution of a complex system. Our weather system is such a complex system. Prevailing weather patterns reflect a steady-state solution to a range of variables, including solar energy, atmospheric density, planetary rotation and axis of rotation influencing prevailing wind patterns, surface and water temperature. A hurricane represents a discontinuity in these systems. It is a departure from the predictable weather patterns found in more stable portions of the logistics curve.

In order to dramatize this point, it has been suggested that theoretically the flapping of a butterfly's wings in the southern hemisphere may result in a hurricane in the northern hemisphere. This sounds far-fetched until we look more closely at the concept of a discontinuity. The formation of a hurricane represents a discontinuity in normal weather patterns. At the point of discontinuity, the system may evolve in dramatically different directions—normal weather patterns versus the development of a hurricane. Theoretically, at that point, even a minor stimulus, such as the air movement of a flapping butterfly wing, may shift the system in one direction rather than the other.[56]

These concepts have relevance to brain function. Day-to-day brain function is best maintained in the reliably predictable middle portion of the logistics curve. Our bodies are continually working to maintain function in this area of the logistics curve—in other words, to maintain homeostasis. We avoid temperature extremes, dehydration, malnutrition, and so forth. With disruption of homeostasis, discontinuities in day-to-day brain function may be encountered in the form of seizures or other forms of dysfunction.

As we develop during childhood, patterns of neuron activity evolve that comprise our sense of self. This pattern of neuron activity is influenced by the positive and negative experiences that we encounter in our life cycles and that define us as individuals. Traumatic experiences disrupt the carrying capacity of a child's social environment. The social environment can no longer reliably foster a robust, coherent sense of self when the child is not screened from trauma. One expects, therefore, to find discontinuities in the child's sense of self. Trauma during childhood may result in a chaotic breakdown of the self, referred to as multiple personality disorder. In multiple personality disorder, the self alternates between different personality fragments. An adult subjected to severe trauma may develop post-traumatic stress disorder. Bipolar disorder represents a bifurcation point at which one's affective state tends to fluctuate between mania and depression.

D. JUGGLERY: THE ART OF INCREMENTAL CHANGE

"The basis of government is jugglery. If it works, and lasts, it becomes policy."[57] I stumbled across this quotation in the book titled *The Middle East* by Bernard Lewis. It is attributed to a ninth-century chief minister in Baghdad. I'd like to focus on this expression for a moment because it captures many of the features of complex systems.

The world we live in is the product of trial and error on a massive scale. Living organisms, for example, come and go. Our world is occupied by the current survivors of this process that Charles Darwin referred to as natural selection. Fortunately, the earth occupies a sufficiently stable orbit about the sun to permit the eons of time required for the trial and error process of biological evolution to take place.

What works and lasts is a form of steady state. There is nothing absolute regarding the specific steady states that evolve. Language formation, for example, represents a form of steady state. The language in use at any time is a product of historical accident and will continue to change over time. The same is true of our governmental systems, as pointed out by a ninth-century chief minister in Baghdad.

Although social evolution proceeds by trial and error, societies have a strong tendency to regard their own language, governmental organization, and religious beliefs as a reflection of absolute truth. In the past, viewing these systems in this fashion has significantly reduced anxiety that is associated with the uncertainties of human existence. In today's world, such absolute confidence in our social systems presents a threat as different social groups are increasingly interacting with one another.

E. INFORMATION TRANSFER

A hurricane has a birth, life, and death. The air molecules contributing the action of the hurricane are continually being modified as they enter and leave the hurricane's pattern of activity. Over time, the hurricane no longer exists as its energy dissipates. All complex systems are time-limited.

Living organisms have a similar life cycle. They require an

ongoing source of energy. Their component molecules are being continually replaced. The molecules of our bodies are replaced several times during the course of a lifetime. Without food, an energy source, and water, upon which the components of our bodies rely to maintain their dynamic patterns of activity, we die. Even with the best of care, we haven't found a way to avoid death.

However, biological organisms are able to pass on a master plan for development through their DNA. DNA is an abbreviation for deoxyribonucleic acid. DNA is a stable molecule that can be passed from one generation to the next and serves as a guide for the development of a next generation.

DNA forms the genes of each cell. Human chromosomes contain a virtual library of genes. Genes provide templates for protein production. They provide instructions regarding the order in which amino acids are strung together to form proteins. Proteins guide the development of the organism cell by cell from inception.

Modification of an organism's genetic inheritance results in a modification of the proteins that shape an organism's development. Approximately 99 percent of the genetic makeup of humans and chimpanzees is identical.[58] The one percent that is different results in a completely different species.

Information transfer through shared language has been critical for evolution of human society. Recall the biblical story about the Tower of Babel. Initially, a group of people working in a very coordinated fashion threatened to build a tower to heaven. This process came to a standstill when they could no longer speak the same language.

Information transfer is a key factor in both biological and social evolution. Biological and social systems take what has worked best in the past and pass it on. Over time, this process leads to an evolutionary transformation of biological and social systems.

Biological systems evolve slowly over many generations. DNA molecules are passed from one generation to another. The vast majority of species rely upon sexual transmission of DNA in chromosomes—long helical strands of DNA. Humans have forty-six chromosomes—twenty three chromosomes from each parent. Sexual transmission of DNA permits increased diversity in the genome over time. Each generation presents their genome (with variations from one person to another) to the threats and opportunities of the earth's biosphere. Those variations that prosper and go on to reproduce survive. Others do not.

Human social systems, relying upon language for information transfer, are able to transfer information much more rapidly. Language permits coordinated interaction among increasingly larger groups of people. Language also enables human beings to transmit information from one generation to the next.

Our facility with language and our survival as a species have been made possible by the speed with which the human brain processes information. Compared to biological and social processes, brain processes are nearly instantaneous. The brain's processing has to be extremely rapid for us to engage in even simple actions that we take for granted. Catching a Frisbee, for example, requires very rapid brain processing. Likewise, language processing requires extremely rapid brain processing.

The rapid processing of information by the human brain is made possible by neural network oscillation and parallel processing. The speed of neural network oscillation determines the limits of environmental activity we are capable of perceiving. Parallel processing permits the brain to create our experience of reality almost instantaneously, permitting our prefrontal lobe to oversee a real-time response to threats.

Our oscillating brain has enabled us to survive as a species. However, failure to recognize the inherent limitations of our brains now threatens our survival. It's essential that we strive for a humbler (more accurate) appreciation of our limitations and a more flexible effort to understand one another. Valuing one another is critical to continued human survival.

CLOSING ————————————————————————

Repetitive activity underlies brain function. Repetitive activity underlies the stability of every complex system. Our planetary system remains stable because the planets return to their starting points with each orbit around the sun. Brain activity is sustained by recurring oscillatory activity within the neural network.

The repetitive activity of the brain's oscillatory loops, referred to as alpha rhythm, underlies all baseline brain function. These oscillatory patterns progressively encompass all areas of the developing brain. The result is a three-dimensional, oscillating neural network.

Conscious experience is associated with gamma wave activity. Sensory stimulation raises baseline alpha wave activity to areas of localized beta wave activity. Integration of prefrontal cortical activity with the multisensory association area increases the frequency of oscillatory activity, resulting in gamma wave activity. Gamma wave activity is widespread involving both motor and sensory cortical areas.

The overall neural network is influenced early on and throughout life by slower-acting second-messenger neurotransmitters produced by neurons of subcortical nuclei. With cortical-cortical connectivity, NMDA neurotransmission permits the formation of declarative memory. Declarative memory provides the cognitive links for the narratives we use to define ourselves. Both slower-acting second-messenger and NMDA neurotransmission contribute to sense of self, our sense of good and bad, and our social interaction. Our sense of self and experience of reality represent a steady-state balance between our developing brains and our experience of the world around us.

Our sense of self is dependent upon the social environment in which we are raised. Individuals from different cultural backgrounds will perceive the world differently. This does not mean that one viewpoint is right and the other wrong, although we have a tendency to divide the world into this type of dichotomy due to the way our brains organize positive and negative information.

Our sense of self is strongly influenced by our positive experiences. A strong sense of self enables an individual to deal constructively with the challenges encountered during a lifetime. The real issue is whether our sense of self works for us in a social environment or interferes with social interaction. Ideally, our sense of self will be balanced by a respect for others.

Social cohesion underlies every society. Society works best if we value one another. It's not an accident that every major religion advocates that we value one another. However, there's a strong tendency to value more those with similar religious beliefs. Valuing one another is critical to the sustainment of social systems. Devaluing one another results in conflict, even chaos.

Social cohesion underlies the success of American society. American society is built upon our recognition that we are created equal regardless of race, sex, religious preference, or national background. This principle has been critical to our success as a society. It has enabled the diverse elements of American society to function synergistically in conditions that could easily have resulted in social discord.

The challenge of maintaining social cohesion in a world made increasingly interactive by advances in communication and transportation technology is particularly important today. How do you maintain social cohesiveness in the face of different social groups jostling for the limited elbow space on this planet? Our survival as a species requires that we learn to embrace diversity rather than expending our energy struggling to eliminate it. None of us have a lock on reality.

NOTES ————————————————

1   D. Purves, *Brains: How They Seem to Work* (New Jersey, USA: Upper Saddle River, 2010), 222.

2   Ibid., 220.

3   MF Bear, GW Connors, MA Paradiso, *Neuroscience Exploring the Brain*, 2nd ed. (Baltimore, USA: Lippincott Williams & Wilkins, 2001), 150–151.

4   M Pertea and SL Salzberg, "Between a Chicken and a Grape: Estimating the Number of Human Genes," *Genome Biology* 11 (May 5, 2010): 206.

5   BJ Baars and NM Gage, *Cognition, Brain, and Consciousness: Introduction to Cognitive Neuroscience* (New York, USA: Academic Press, 2007), 455.

6   Ibid., 7.

7   P Rakic and PJ Lombroso, "Development of the Cerebral Cortex: I. Forming the Cortical Structure," *Journal of American Academy of Child and Adolescent Psychiatry* 37, no. 1 (Jan. 1998): 116–117.

8   PJ Lombroso, "Development of the Cerebral Cortex: VI. Growth Factors: I.," *Journal of American Academy of Child and Adolescent Psychiatry* 37, no. 6 (June 1998): 674–675.

9   F Vaccarino and PJ Lombroso, "Development of the Cerebral Cortex: VII. Growth Factors: II.," *Journal of American Academy of Child and Adolescent Psychiatry* 37, no. 7 (July 1998): 789–790.

10 F Conti, "Localization of NMDA Receptors in the Cerebral Cortex: a schematic overview," *Brazilian Journal of Medical and Biological Research* 30, no. 5 (May 1997): 555–560.

11 VB Mountcastle, "The Columnar Organization of the Neocortex," *Brain* 120 (1997): 701–722.

12 Baars and Gage, *Cognition*, 327.

13 N Anderson, "Scientist's Solve Mystery of Human Brain's Chandelier Cells," Sci-News.com, Nov. 27, 2012.

14 G Randhawa, "How Chandelier Cells Light Up Human Thought," *New Scientist* 14, no. 51 (Sep. 3, 2008).

15 N Wong, "Theories of Personality and Psychopathology," in *Comprehensive Textbook of Psychiatry/V*, 5th ed., edited by HI Kaplan and BJ Sadock (Baltimore, USA: Williams & Wilkins, 1986), 382–383.

16 ER Kandel, "Cellular Mechanisms of Learning and the Biological Basis of Individuality," in *Principles of Neural Science*, 4th ed., edited by ER Kandel, JH Schwartz, and TM Jessell (New York, USA: McGraw-Hill, 2000), 1260.

17 A Fisahn, FG Pike, EB Buhl, and O Paulsen, "Cholinergic Induction of Network Oscillations at 40 Hz in the Hippocampus in Vitro," *Nature* 394 (July 9, 1998): 186–192.

18 Bear, Connors, and Paradiso, *Neuroscience*, 150–151.

19 ER Kandel and SA Siegelbaum, "Synaptic Integration," in *Principles of Neural Science*, 4th ed., edited by ER Kandel, JH Schwartz, and TM Jessell (New York, USA: McGraw-Hill, 2000), 207–228.

20 ER Kandel, TM Jessell, and JR Sanes, "Sensory Experience and the Fine-Tuning of Synaptic Connections," in *Principles of Neural Science*, 4th ed., edited by ER Kandel, JH Schwartz, and TM Jessell (New York, USA: McGraw-Hill, 2000), 1115–1129.

21 Fisahn, Pike, Buhl, and Paulsen, "Cholinergic Induction," 186–188.

22 C Fine (introduction by), *The Britannica Guide to the Brain* (Philadelphia, USA: Running Book Press Publishers, 2008), 160–162.

23 F Crick, *The Astonishing Hypothesis: The Scientific Search for the Soul* (New York, USA: Simon & Schuster, 1994), 243–268.

24 L Sanders, "The Probabilistic Mind Human Brains Evolved to Deal with Doubt," *Science News* (Oct. 8, 2011): 18–25.

25 J LeDoux, *Synaptic Self: How Our Brains Become Who We Are* (New York, USA: Penguin Putman Inc., 2002), 246–258.

26 Wong, "Theories of Personality," 356.

27 Ibid., 362.

28 S Freud, *Beyond the Pleasure Principle* (New York, USA: Norton & Company, Inc, 1961), 1–5.

29 HI Kaplan and BJ Sadock, *Comprehensive Textbook of Psychiatry V*, fifth edition (Baltimore, USA: Williams and Wilkins, 1989), 428.

30 PD Kramer, *Listening to Prozac* (New York, USA: Penguin Books, 1993), 67–77, 87–107.

31 D McNiel and P Freiberger, *Fuzzy Logic* (New York, USA: Touchstone, 1994).

32 Ibid., 230–231.

33 Ibid., 38–40.

34 T Kuhn, *The Structure of Scientific Revolutions* (Chicago: University of Chicago Press, 1962).

35 McNiel and Freiberger, *Fuzzy Logic*, 230.

36 C Darwin, *The Origin of Species* (Franklin Center, USA: Franklin Library, 1957), 70.

37 E Mayr, *Toward a New Philosophy of Biology* (Cambridge, USA: Harvard University Press, 1988), 172–173.

38 Johns Hopkins Medical Institutions, "This Is Your Brain On Jazz: Researchers Use MRI To Study Spontaneity, Creativity," *Science Daily* (Feb. 28, 2008), retrieved Sep. 3, 2011, http://www.sciencedaily.com/releases/2008/02/080226213431.htm.

39 DA Gunsard et al., "Medial Prefrontal Cortex and Self-Referential Activity: Relation to a Default Mode of Brain Function," published online before print, March 27, 2001, www.pnas.org/content/98/7/4259.full.

40 CG Jung, *Man and His Symbols* (New York, USA: Dell Publishing Company, Inc., 1975), 25–26.

41 M Buber, *I and Thou* (New York, USA: Touchstone, 1996), 51–56.

42 CG Jung, *Analytical Psychology: Its Theory and Practice* (New York, USA: Vintage Books, 1968), 7.

43 Ibid.

44 Ibid., 10.

45 Ibid.

46 Ibid.

47 Ibid., 40.

48 Ibid., 41.

49 M Hines, P, "Prenatal Gonadal Hormone and Sex Differences in Human Behavior," *Psychological Bulletin*, (1982, Vol 92 No 1): 56-80.

50 R Restak, *The Brain: The Last Frontier* (New York, USA: Doubleday & Co. Inc., 1979), 198-204.

51 The story of Hurricane Katrina is taken from the Wikipedia website, http://en.wikipedia.org/wiki/Hurricane_Katrina.

52 J Gleick, *Chaos, Making a New Science* (New York, USA: Penguin Books, 1987), 121–153.

53 EP Odum, *Basic Ecology* (Philadelphia, USA: Saunders College Publishing, 1983), 156–159.

54 Ibid., 158.

55 Gleick, *Chaos*, 59–80.

56 Ibid., 11–31.

57 B Lewis, *The Middle East* (New York, USA: Scribner, 1995), 155.

58 J Reader, *A Biography of the Continent Africa* (New York, USA: Vintage Books, 1997), 78–79.